T0251484

Procurement of Built Assets

Duncan Cartlidge

Spon Press
an imprint of Taylor & Francis
LONDON AND NEW YORK

First published by Butterworth-Heinemann

This edition published 2011 by Spon Press
2 Park Square, Milton Park, Abingdon, Oxon OX14 4RN

Simultaneously published in the USA and Canada
by Taylor & Francis Group, 711 Third Avenue, New York, NY 10017, USA

Spon Press is an imprint of the Taylor & Francis Group, an informa business

First published 2004

British Library Cataloguing in Publication Data
Cartlidge, Duncan P.
    Procurement of built assets
    1. Construction industry – Management 2. Industrial
procurement 3. Risk management
    I. Title
    624′.0687

    ISBN 07506 58193

Library of Congress Cataloging in Publication Data
Cartlidge, Duncan P.
    Procurement of built assets / Duncan Cartlidge.
        p. cm.
    Includes index.
    ISBN 0-7506-5819-3
        1. Construction industry. 2. Industrial procurement.
        I. Title.

HD9715.A2C355 2004
624′.068′7–dc22   2003049434

Contents

List of figures and tables

Foreword

It is only in the last three decades or so that procurement has been regarded by business as a strategic issue, rather than as an incidental activity to the real processes of production and marketing. Within organizations purchasing tended to be where you put people who had no professional qualifications, or the juniors in the organization. After all, anybody can buy; we do it every day!

All this changed in the 1970s and 1980s, when the Japanese in particular, showed how integrating the procurement function into the supply chain paid handsome dividends in controlling quality, production processes, and whole life costs. The introduction of just-in-time techniques into manufacturing and retail industries made project managers, and Financial Directors, identify and assess the true costs of holding stock, inefficient site control, and sub-optimal assembly programmes. The relatively high profitability of UK retailing during the 1990s gave ample evidence of the benefits resulting. Procurement, as a strategic discipline in the UK economy, had arrived.

Unfortunately, the UK construction industry was somewhat tardy in receiving the new gospel. On the supply side there seemed to be a conviction, sometimes verging on complacency, that, since most of the technical expertise rested with them, the customer would be well advised to take the product they offered and be thankful. At least Henry Ford assured the consumer they could have any colour they liked as long as it was black; the supply side of the construction industry seemed to offer any colour the supply side liked, immaterial of the customer desire. Clients, on the other hand, seemed to think that the new procurement techniques that they used to benefit their core business had no relevance when it came to acquiring their construction needs. Or at least that's the advice they received from construction professionals.

Unfair? Perhaps so; but it was this uneasy feeling that the construction client, whether in the public or the private sector, was not getting a fair deal that has led to the unprecedented self-examination that the industry has undertaken over the last few years. Governments of both political persuasions recognized that, if they were to meet the aspirations of society for better services and higher living standards, a reconstruction of the obsolete infrastructure and provision of an improved built environment had to be provided. Private industry too recognized the need to meet changing technologies and the concentration in the developed world on the service industries. This in turn generated major construction programmes of new office building and adaptation, and relocation of existing production capacity. By 1998, home demand on the UK construction industry accounted for some £58 billions annually (including DIY), or nearly 5 per cent of UK GDP. So, when the Government inspired examinations of the industry, first by Sir Michael Latham and then by Sir John Egan, confirmed that there was scope for a potential 30 per cent improvement in the industry's productivity, there was a strong incentive to break from the outdated and inefficient methods, and from the long-established positions based on the preservation of special interests.

I have been fortunate enough to be involved in what has been regarded as an internally directed revolution in the industry, both from the suppliers' and the clients' points of view. The strongest impression that has been left with me is how markedly the language and culture of the industry is changing towards the acceptance of a fully integrated team approach in identifying clients' construction needs and satisfying them in ways which add value to their core business. This collaborative approach, the recognized basis for procurement strategies in other industries, has been the key element in the recommendations for making the industry truly effective in providing construction solutions for clients' business requirements. It is concerned with obtaining best value for money over the whole life of the project, and has been promulgated through the industry's strategic machinery, now being carried forward in the industry's Strategic Forum, which seeks to bring together all participants in the process, including the clients. It would be wrong to depict this changing business environment in construction as being insufficiently 'hard-nosed' when seen in the historic context of the industry's confrontational way of doing business. The fact is that other industrial sectors in the UK economy have accepted for a long time that better solutions emerge from the collaborative input of all the specialists involved in the production process and including the end users. Rather than from an arrogant assumption that the customer is unable to articulate what their true requirements are, and will therefore have to accept what their suppliers decide to give them. We are now seeing from the industry schemes, which incorporate genuinely valuable contributions towards the resolution of complex technical problems not only from designers and contractors, but also from specialist suppliers, installers and manufacturers, as well as from the users of the project, themselves. It is of course the client who pays the bills for the project, not just in terms of the capital costs of the works, but also the costs of operation and maintenance of the project that they need in order to be able to undertake their business. The input to the process, and the understanding of it on the part of their suppliers, of the definition of the client's total requirements should lead to the provision of a facility which enhances the client's position and gives them a genuine asset which adds value to their core business. By concentrating on the integration of the process encourages, I believe, a genuinely holistic approach to solving the client's business problems. This in turn recognizes that corporate social responsibility, environmental protection and enhancement, and sustainability are intrinsically bound up in the final project. Construction projects, more than most other economic products, are comparatively long lasting and have an immediately discernible effect on a wider public than those for whom they are specifically provided. The best construction projects today incorporate quantified assessments of whole life performance and costs, as well as sensitivity in the choice of materials and design, giving due importance to the availability and use of renewable resources, and the retention of valuable professional and craft skills.

It will be readily recognized from what I have said above, that one of the most important contributions that can be made towards the achievement of the 'modern way' in construction is the adoption of a truly integrated procurement process, utilizing the various skills, including those of the client, in a genuine partnership with shared aims and rewards. This book emphasizes precisely that, and illustrates how important it is to give up once and for all the old adversarialism, particularly when problems arise.

I am happy to commend this book to a wide representation of the industry.

Zara Lamont OBE

Preface

Almost 30 years ago I sat down to write my first book; it was called *Cost Planning and Building Economics* and included by way of a dedication, the following quotation, attributed to the Egyptian King, Cheops, at the commencement of the building of the Great Pyramid of Khufu in 2589 BC,

I don't care how much it costs or how long it takes.

It seemed appropriate at the time to include these words as a sort of wake-up call to the UK construction industry, who had the unenviable reputation of delivering the majority of its products, over budget and over time. Thirty years, five books and several centuries later than Cheops's quote, yet another obelisk was being constructed; the home of the newly devolved Scottish Parliament in Edinburgh, Scotland. This particular project is referred to several times within the book, and while I would not wish readers to conclude that I am conducting a personal vendetta against this project – it is something of a horror story! Imagine going into your local car dealer to collect your new car to be told; sorry but the price has increased by over 90 per cent and it will not be ready for at least another year. Here is the deal – take it or leave it!

Utilizing management contracting and heavily criticized because of costs that have escalated from the original estimate of £40 millions to more than an estimated £375 millions, the Scottish Members of Parliament who have been tasked with overseeing the Parliament's construction, decided to visit the Aberdeenshire quarry where the stone that is to clad the new building is being cut. A young reporter from a local television station, seized the moment, thrusting a microphone towards an SMP he asked the question of the delegation, '*aren't you concerned about the criticism of escalating costs and incidentally, what is the final bill?*' The SMP leading the group looked astonished and replied '*This project is being procured using competitive tendering and it is impossible to know the final cost until all the tenders have been received and the job is complete.*' and with this he waved the journalist away to continue the tour. How little seems to have changed when it comes to procurement of buildings during the past 4700 years, Cheops's quote could equally well be applied to this project.

A further indication of attitudes to construction procurement was given in an article in the *Sunday Telegraph* (25 April 2003) concerning the same 'Edinburgh Extravagance' and quoted the editor of the *Journal of Scottish Architecture* who upon hearing that contractors on the project had been paid advances of £250,000 replied '*it is unheard of for cash to be advanced to contractors in this way*'. Not so, in France many contractors, including the really big international names, ask for advances at the start of a project as part of normal business, a practice that will be discussed in Chapter 8. You see in France, the client actually trusts the contractor not to disappear off on the next airplane with the cash, like some sort of criminal. As a final word on the Scottish Parliament building, well for now anyway, perhaps it is not entirely fair to judge the performance of the UK construction industry from high profile mega projects, as they appear to have their own particular multi-layered agendas, as will be discussed in Chapter 2.

This book draws heavily on the experiences and best practice of other industries and market sectors who have, just as construction is now having to do, taken a critical look at their procurement practices and techniques and the inherent waste in many traditional systems. From his experience in education it has become apparent to the author, that in a majority of cases, architects are taught to approach procurement from the elitist stance that, the process is 'architect led' and by definition everybody else follows: commercial managers from the point of exploiting an inherently flawed process to maximize profit for the contractor, including a good measure of bullying sub-contractors and quantity surveyors approach the process from the starting point of – 'let the wars begin.' Rarely does the client seem to be the main focus.

For the majority of construction professionals, the subject of procurement is most often introduced and taught along the stereotypes referred to in the previous paragraph. Consequently, when it comes to the choice of a procurement path few construction professionals know the basics of, for example; good procurement, supply chain management, value for money, or how to select an appropriate strategy. Organizations in all market sectors increasingly operate in a service-oriented culture where the expectations of consumers and indeed construction clients, have been inevitably raised. In general, the approach to the construction procurement process has not yet moved to a consumer orientated approach, although, there has been increasing acknowledgement that customer care must become part of the provision of built assets and as such will be a major step in achieving the aims of Latham and Egan and the many initiative that these reports spawned.

There can be no doubt – the creation of a complex new building carries inherent risks, both technically and commercially and improving the procurement performance is about changing the hearts and minds of professionals, of whatever discipline, involved with the procurement of built assets, preferably at the start of their career.

Quite simply therefore, the rationale of this book, is to introduce best practice construction procurement to a wide range of construction professionals, clients and the industry as a whole in the expectation that construction procurement becomes more responsive to clients' needs and helps deliver projects on time and to budget.

Duncan Cartlidge
www.duncancartlidge.co.uk

Acknowledgements

My sincere thanks go to the following people who have contributed their enthusiasm, time and considerable expertise to
the contents of this book.

Kevin Thomas for case study 2, Chapter 7

Kevin Thomas is Director of Worldwide Strategic Planning for
GlaxoSmithKline Research and Development. With a background in the defence and pharmaceutical industries, Kevin
has over 25 years' experience in the construction, operation
and reconfiguration of buildings as well as sites and the management of the facilities and services provided to their occupants. Kevin was instrumental in redefining Glaxo Wellcome's
approach to R&D construction activity as sponsor of its Fusion
programme. Fusion was successfully applied to the restructuring of the company's Biopharmaceuticals site at Beckenham,
Kent, winning the 1999 Contract Journal Single Project
Partnering Award.

Kevin is Deputy Chairman of 'Collaborating for the Built
Environment' and has been a lead member of the Strategic
Forum Integrated Teams Working Group who have been draft-
ing the Strategic Forum Toolkit.

Rory Lamont for case study 1, Chapter 7

Rory Lamont is a Supply Chain Management Specialist for
British Petroleum's UK Continental Shelf Operations. As well
as championing the appropriate use of reverse auctions in the

upstream segment of the business, he also looks after BP's engineering consultancy, equipment condition monitoring and personal protective equipment contracts for their North Sea operations. His specific area of interest is inter-company relationships and the role played therein by technology. lamontr@bp.com.

Sean Lockie for advice, comments and contribution to Chapter 3

Sean Lockie is a technical consultant who is part of the Whole Life Value division at Atkins Faithfull and Gould. He has over 10 years' experience in the property sector covering a range of projects in commercial, educational, leisure and health sectors. He has worked on over 30 PPP projects.

Faithfull and Gould are part of the Atkins Group of Companies. Atkins is one of the world's leading providers of professional, technology-based consultancy and support services, with over 14,000 staff worldwide.

Mohamed Kishk, The Robert Gordon University, for his contribution on whole life costs in Chapter 3. Since his return to academia in 1998 to start a PhD programme in whole life costing based decision-making Mohamed has published more than 25 articles on WLC in referred academic journals as well as giving presentations at major regional, national and international conferences.

Cliff Jones, NHS Estates, Leeds, for his advice and guidance on NHS ProCure 21.

Ray Robinson, AON Risk Services, London for his advice on insurance matters.

Dedication

John (Jack) Cartlidge 1914–1980

1

Construction procurement – the case for a new way

In no other important industry is the responsibility for the design so far removed from the responsibilities of production.
—Sir Harold Emerson 1964

Introduction

The ability successfully to procure built assets is at the heart of the construction process and in turn at the heart of the procurement process is identifying the constantly evolving needs of the construction client.

Just what every construction client wants!

What are the drivers of construction procurement in the UK and are they so much different to other sectors? If one were to ask some of the 2.5 million customers in the UK who annually buy a new car – what were the criteria for buying a particular model? – the answers would probably be:

- long warranties, reliability, no defects;
- fitness for purpose, that is, 4 × 4, family saloon, etc.;
- features, such as air conditioning, CD player included as standard;
- long service intervals and low running and maintenance costs;
- delivered promptly.

In fact criteria that all in all, add up to perceived value for money over the life span of the vehicle. Criteria that appear to be

Figure 1.1 The construction client's dilemma.

recognized and taken into consideration by the automotive industry, as the following quote from BMW's marketing information illustrates:

> *The true cost of the vehicles on your fleet is a balanced measure of all the relevant attributable costs. Servicing, on-the-road price, depreciation and days off road are all essential measures that will directly affect your fleet budget. With high residual values, lengthy servicing intervals and excellent build quality, the BMW and MINI range are well placed to offer some significant advantages over competitive marques … with the Whole Life Costs of the BMW and MINI ranges you will be pleasantly surprised to find that it makes financial sense to offer our cars to your company car drivers as part of your fleet.*

Admittedly the life span of a family car is not as long as the average new construction project, but if only procuring built assets sounded so simple. As illustrated in Figure 1.1 not only do construction clients traditionally have to grapple with a

Table 1.1 Key performance indicators

Key performance indicator	Measure	2000	2001
Client satisfaction			
Product	Scoring 8/10 or better	72%	73%
Service	Scoring 8/10 or better	63%	65%
Defects	Scoring 8/10 or better	53%	58%
Safety	Mean accident/incident rate/ 100,000 employed	1088	990
Cost predictability			
Design	On target or better	63%	63%
Construction	On target or better	48%	50%
Project	On target or better	46%	48%
Time predictability			
Design	On target or better	41%	46%
Construction	On target or better	59%	62%
Project	On target or better	36%	42%

Key: 10 = totally satisfied, 5/6 = neither satisfied nor dissatisfied,
1 = totally dissatisfied.
Source: dti (2002), *Construction Statistics Annual*, HMSO.

whole range of diverse issues, but in addition with a diverse range of professional and technical advice.

Similarly, if British Airways were to consider replacing its ageing transatlantic fleet of Boeing 747 jets, the criteria for choosing the most appropriate aircraft would probably be similar to those of the car buyer. By way of contrast, Table 1.1 taken from the dti 2002 annual construction industry performance indicates that 10 years post-Latham the UK construction industry is still failing to meet the expectations of its clients. How many car buyers, for example, would be prepared to accept a 58% rating for defects?

Why then does it appear that construction clients are still unable to find the same levels of value for money as the average car buyer?

The case for a new approach to procurement

The challenge that has been laid at the feet of the construction industry is to transform the diverse and often separate processes

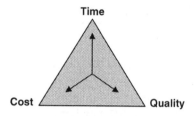

Figure 1.2 Procurement drivers.

of design and procurement of built assets into one single integrated production process.

The long established principal drivers for construction procurement have been said to be (it's no coincidence that the model is pyramid shaped) time, cost and quality as illustrated in Figure 1.2.

Traditionally, the relative importance of these drivers determined the selection of procurement path. The client being asked to identify which of the drivers was the most important – the inference being that it is impossible to have all three simultaneously. Therefore, if time were of the essence then cost and quality would have to be compromised and vice versa. However things have moved on and modern forward thinking construction clients are now taking as read that, just like the car buyer, supply chains can deliver all three of the above, that is: built assets that are delivered on time, within the cost target with zero defects. For example, later in this book the role of frameworks and procurement will be discussed along with the rigorous prequalification processes during which contractors have to demonstrate their ability to deliver added value as a prerequisite to even being placed on a tender list. What now, therefore, do construction clients want from their construction industry partners in the field of procurement? Progressive clients, some of whom have contributed case studies to this book, are now demanding that construction procurement is not detached from their mainstream business operations, organized and delivered by construction professionals operating to their own agendas. Progressive clients are now demanding that procurement mirrors corporate strategies and targets including addressing the following areas:

- Understanding clients' needs which may, depending on circumstances, include producing solutions that satisfy

the following criteria:
- flexible and adaptable facilities,
- maximize use of existing assets,
- immediate start and early finish,
- minimal waste and defects,
- lower and predictable whole life costs,
- greater predictability in cost and time,
- a long-term vision of short-term demands.
- The ability to demonstrate pro-activity and innovation by:
  - questioning and challenging clients' needs,
  - develop the ability to question and challenge conventional practice, on the basis that in many cases so far, it has not produced satisfactory solutions.

Collaboration with clients and supply chains

From the mountain of information, reports and statistics that are available, relating to the performance of the UK construction industry during the past 50 years or so, it would appear that the case for a new approach to procuring built assets has been proven. However, despite all of these studies, spanning both the public and private sectors, the UK construction industry is letting its clients down and is perceived by many other industries to be a dinosaur in procurement terms with each new initiative being met with a barrage of scepticism and resistance.

It has been argued that one reason for construction's comparatively poor performance is that the construction process is unique. This uniqueness is said to be characterized by a wide spectrum of clients as follows:

- Small one-off client or large corporate investor.
- Occasional or experienced client.
- Public or private sector client (Morledge *et al.*, 2002).

Unfortunately, the Luddites within the industry have traditionally seized on these facts to champion the cause of Not In My Industry Thank You! (NIMITY). It could be argued that a similar client/end user spectrum can be found in a number of market sectors, for example computing, but this has not prevented other industries from moving forward and adopting integrated procurement practices. In addition to the above factors the following views have also been identified as peculiarities of construction products and construction (source: adapted from Koskela (2003)):

- The complexity of organization and manufacture of built assets.
- The small extent of the penetration of standardization.
- The high turnover of workers.
- The localized nature of orders of extraordinary diversity.

Why do so many projects fail to meet targets?

Poor procurement performance transcends both the public and private sectors and some clues for this may come from a government sponsored report. In July 2002 Mott MacDonald prepared a *Review of Large Public Procurement in the UK* for HM Treasury, as part of the Green Book Review, referred to in more detail in Chapter 3. One of the areas considered by the review was the poor performance of UK public sector construction projects. The study identified what it called 'high levels of optimism' as one of the principal causes for poor performance. Optimism being defined as 'the tendency to underestimate project costs and duration and overestimate project benefits, or the inability to identify and mitigate risk.' The study continued to identify critical project risk areas that cause cost and time overruns where optimism bias was found. The report calculated an optimism bias, which is expressed as a numerical indication of the level of such bias and can be applied to estimates in order to improve reliability. The report concluded generally that the performance of projects procured using public private partnerships (PPP) was much higher due in part to the much more rigorous approach to the establishment of a robust and realistic business case, combined with rigorous risk analysis. This point is underlined in the case of the new Scottish Parliament project, see Preface, where a similarly complex project, the Edinburgh Royal Infirmary, was completed on time and to budget using a PPP/Private finance initiative (PFI) approach. It is not surprising that the optimism bias levels for PPP/PFI projects are lower than for some traditionally procured projects, as more project risks are identified

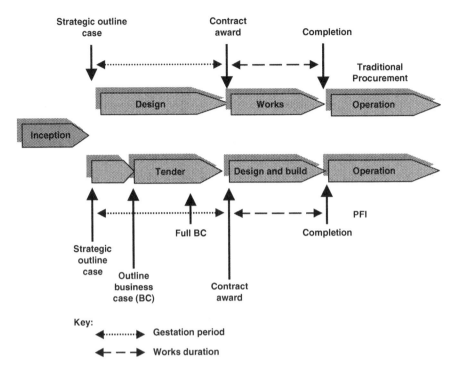

Figure 1.3 Review of large public procurement in the UK (source: Mott MacDonald, 2002).

and mitigated at the full business case stage than at the strategic outline case and the outline business case stages. As illustrated in Figure 1.3 the PFI procurement route contains a much more rigorous procedure for business case development than traditional procurement routes. The review concluded that one of the primary objectives of the business case is to identify risk. The Mott MacDonald report also gives an indication of the project risk areas most likely to cause overruns if sufficient risk mitigation strategies are not in place. It would appear therefore that the provision of a sound business case and the correct identification of risk are of great importance in ensuring efficient project delivery. Both of these topics will be discussed in Chapter 2.

A more controversial explanation as to why large scale public sector projects such as the Channel Tunnel so often are a procurement disaster, is put forward by Bent Flyvbjerg (*Megaprojects and Risk: An Anatomy of Ambition*, 2003). Concluding that bad procurement is not just a public sector phenomenon,

Flyvbjerg and his co-authors at Aalborg University in Demark, Bruzelius and Rothengatter, examined over 250 mega projects, from all over the world, mainly in the transport sector, undertaken between 1924 and 1998. As well as reaching the same conclusion as Mott MacDonald (2002), namely that civil servants have a tendency to be overly optimistic when it comes to estimating costs and time scales, they go further by suggesting that there is often a good deal of deceit at the planning stages of large prestigious projects by politicians and professionals alike: '*Cost underestimation and overruns cannot be explained by error and seem to be best explained by strategic misrepresentation, namely lying, with a view to getting projects started.*' The authors go on to suggest that criminal penalties should be introduced if this type of conduct were found to be responsible for project excesses!

Is construction that unique?

Koskela (1998) rejects the premise that construction is unique and suggests that it is in fact just another production system. According to Koskela the differentiating characteristics often cited by the construction industry of one-of-a-kind nature of projects, site production and temporary teams, are present in several other types of production.

There follows the sequence of events, based on *The Machine that Changed the World*, Womack *et al.*, of the typical approach towards the procurement of a new motor car around 1970/1980:

- The overall concept is planned in detail by the senior management.
- Detailed drawings and specifications are produced for each part, such as steering wheels, bumpers, etc.
- Only at this point are the organizations, the suppliers (typically 1000–2500), who will actually make the parts, called into the process. However, by this time it is too late to improve the design.
- The suppliers are shown the drawings and asked to produce bids.
- The suppliers know from experience that from the assemblers' point of view 'cost comes first'. Therefore, quoting a low price is absolutely essential to winning a bid. This practice

leads to implausible bids winning contracts followed by cost adjustments that eventually make the cost per item higher than those of realistic, but losing bidders.

- Since this is the case should they, the suppliers, bid below cost, because as the suppliers also know, once production has commenced they may be able to go back for cost adjustments and variations.
- The mass production assembler has played this game thousands of times and fully expects the successful suppliers to come back for price adjustments.
- The assemblers would dearly like to know the suppliers' real costs. But these are jealously guarded by the suppliers in the belief that by revealing only the price per component, they are maximizing their ability to hide their true profits from the assembler. Playing the bidders off makes them very reluctant to share ideas on improved production techniques.
- Once the suppliers have been chosen they start to produce the parts. Often, many problems are uncovered as the components are produced by suppliers who have no direct contact with each other.
- Revised drawings are produced.
- The new model reaches the market often to find that something is not right – the cars have to be brought in for modifications.

For those familiar with the traditional approach to the procurement of built assets the above sequence will be all too familiar, and some reassurance can be taken from the fact that adversarial practices have not been confined to the construction industry!

Despite the apparent similarities in the procurement of motor cars and built assets there is no disputing or overlooking the fact that construction is a very diverse activity, operating at a variety of levels, from complex civil engineering projects to simple domestic alterations and consequently procurement will require an equally diverse approach. Also it should be remembered that despite the rhetoric large organizations continue to run many different procurement methods, therefore it is important to ask questions each time, for each new project. Problems may arise if, when for example, organizations decide to adopt a single approach to procurement as has happened in some public sector agencies in the UK. Although it must be stated that it is not always necessarily a poor strategy to adopt this approach. For example, National Health Service projects in England and Wales are very unlikely to proceed unless the Private Finance Initiative or ProCure 21 procurement paths are used and both of these systems will be discussed in Chapters 4 and 5. Even if the uniqueness of construction is accepted the industry has proved to be less able to develop radical new ways of working which build on the experience of other industries. In addition, the UK construction industry, with a turnover of £70 billion per annum has a particularly unique structure.

According to the Department of the Environment, Trade and the Regions (detr) in 2001, out of a total of approximately 168,000 firms and a total workforce of approximately 1.5 million, in the UK construction industry less than 90 firms or 0.5% have a work force of over 600, while approximately 130,000 firms or 72% employ between 1 and 3 people, with over half the total number of firms being single person organizations. However, the top 0.5% of companies carries out approximately 20% of all construction work in terms of value. This structure is by no means unique on the world stage and in Chapter 8 it can be seen that figures for the Australian construction industry reveal similar patterns. By comparison, the automotive industry has a turnover of £40 billion per annum and employs approximately 800,000 people but has only 10 major manufacturers. The consequences of this structure for UK construction are as follows:

- Difficulties in developing the 'visionary leadership' called for in the many government initiatives such as Accelerating Change and in effecting widespread change, as there are few large organizations able to champion new ideas and approaches.
- Difficulty in introducing new skills to the 130,000 firms who do not have the resources, time or inclination to become familiar with techniques such as e-procurement and supply chain management.
- New initiatives are perceived as being relevant for the top companies only, the so-called 'A list' leaving the majority of the industry to maintain the *status quo*.
- A paltry £147 million per annum spent on research and development in an uncoordinated and unfocused way.

Table 1.2 is Koskela's overview on problems related to the peculiarities of construction procurement with corresponding solutions. In this table process control refers to the management of a project, process improvement to the development efforts of the permanent organizations in construction (designing, manufacturing of materials and components, and contracting).

One of the fields where the UK construction industry does seem to lead the field is silo working, that is working in isolation, guarding and protecting any information which it is feared may give commercial advantage to a competitor. In this approach the stakeholders in the construction supply chain work in isolation, that is, the design team keep the contractors and they in turn keep the sub-contractors at arms length for as long as possible into the project, and as a result a significant proportion of built assets are procured by means of a system where the design and production are separate operations and lack buildability. As will be discussed later in the chapter the breaking down of this silo mentality is a major step in moving away from current construction procurement culture. One of the characteristics of the many new approaches to procurement is the abolition of fear and the creation of a safe commercial environment in which suppliers can operate.

A prime example of the benefits of how project cooperation can be of benefit comes from another sector. The Human Genome Project, started in 1990, was originally a 13-year effort to identify all of the approximately 30,000 genes in human DNA and to store this information on databases. The importance of this project to mankind is self-evident, as is the tremendous commercial importance of the research to a fiercely competitive pharmaceutical industry. If construction is considered to be a risky business, then consider the development of a new drug; £250–300 million in development costs over a 10-year period before any return and a patent limited to 20 years for successful products. The outcome of the Genome project therefore is clearly of great interest to commercial biotechnical organizations worldwide, searching for the illusive next blockbuster drug. However, the project was unique in one very important way; it was a collaborative exercise, with the transfer of the technologies used in the research and the subsequent results, to the private sector, freely available without constraints, via the Internet. Originally, the project was planned to last for 15 years but collaborative working has shortened the

Table 1.2 Application of the new production philosophy to construction

Peculiarity of construction	Process control problems	Process improvement problems	Structural solutions	Operational solutions for control	Operational solutions for improvement
One-of-a-kind	No prototype cycles. Unsystematic client input. Coordination of uncertain activities	Processes do not repeat – long-term improvement questionable	Minimize the one-of-a-kind content in the project	Upfront requirements analysis. Set up artificial cycles. Buffer uncertain tasks	Enhance flexibility to cover a wider range of projects. Accumulate feedback
Site production	External uncertainties: weather, etc. Internal uncertainties and complexities: flow interdependencies, changing layout, viability of manual work	Difficulty of transferring improvement across sites solely in procedures and skills	Minimize the activities on site in any material flow	Detailed and continuous planning. Multi-skilled work teams	Enhance planning and risk analysis capability
Temporary organization	Internal uncertainties: exchange of information across organization borders	Difficulties of stimulating and accumulating improvement across organization borders	Minimize temporary organizational interfaces	Team building during the project	Integrate flows through partnerships

Source: Koskela (1992).

programme by 2 years than otherwise would be possible with silo working, thereby potentially saving thousands of lives as the findings are used in worldwide research projects. Full details of the human genome project are at www.ensembl.org/

A genealogy of procurement

As previously discussed, the UK construction industry, unlike other major industrial sectors, has never appeared to have seen the necessity to challenge long held industry beliefs/practice or to question the *status quo*. In the main, change has been brought about as the result of client pressure or economic circumstances or a combination of both, but seldom from within the industry itself. The 21st century is witnessing some construction clients developing the same expectations that, for example, retail customers now regard as normal. As Table 1.1 illustrates construction still has a long way to go before it matches, for example, the defects target of five parts per million demanded by Toyota. Figure 1.4 plots the correlation of market trends with the development and use of the predominant procurement strategies demanded by clients.

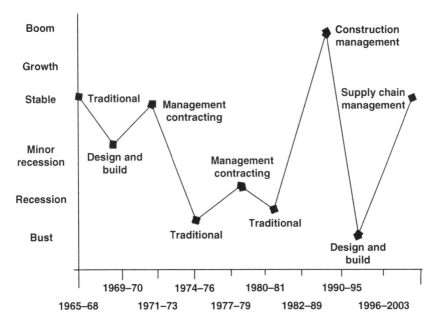

Figure 1.4 Economic and procurement trends (source: Gardiner and Theobald 2002).

The UK construction industry has always been subject to periods of boom and bust. Traditionally the quiet periods have been times of reflection for the industry and the publication of reports cataloguing the deficiencies of the system. Organizations vow to improve performance, promises that usually fall by the way side as soon as the flickers of an upturn in the market appear on the horizon.

The property market has always driven procurement trends and not vice versa. During the past 160 years or so, there have been a wide variety of fashionable procurement strategies adopted by the construction industry. Table 1.3, a genealogy of procurement, attempts to relate the emergence of new procurement paths to external drivers.

1834–1945: the quantity surveyor and contracts

Long before Cheops commissioned the building of the great pyramid, clients had been procuring built assets. However, this brief history of construction procurement will start in 1834 – notable for the birth of the quantity surveyor, for most people a profession still synonymous with construction procurement. The reason behind the quantity surveyor's emergence at this time was architect's desire to rid their professional institution of surveyors and their perceived obnoxious commercial interest in construction. Also during this period the emergence of another major influence on procurement was The Standard Form of Contract. The Joint Contracts Tribunal in their publication The Use of Standard Forms of Building Contract advises that from about 1870 the possibility of a standard form was discussed among various trade bodies and the Royal Institute of British Architects (RIBA). Agreed forms were subsequently issued in 1909 and 1931. New editions followed to take account of change in industry practice.

1946–1969: post-war regeneration

For the UK construction industry, the legacy of the Second World War was a massive demand for new buildings and infrastructure. Not only had many major cities, like London, Portsmouth and Liverpool suffered extensive bombing damage during the war, but in addition the 1950s and 1960s saw vast tracts of inner

Table 1.3 A genealogy of procurement

Dates	Economic milestones	Procurement trends	Construction activity
1834–1945	Few corporate clients	Sequential, fragmented process Bills of quantities, competitive tendering	Traditional approach
1946–1969	Post-war regeneration	High value = low cost Lump sum competitive tendering Cost reimbursement	Rebuilding post-war Britain
1970–1979	Rampant inflation 25% + pa Historically high interest rates	Management contracting Two stage tendering	Property boom 1970–1974
1980–1989	1989 base rate reaches 15% Financial deregulation Privatization 1987 inflation reaches 7.7% pa 1987 stock market crash	Construction management Management contracting Compulsary Competitive Tendering Bespoke contracts to load risk onto contractors	Property slump 1980–1984 Property boom 1985–1990
1990–2000	Globalization Low interest rates and inflation World economic slump	Partnering PPP/PFI	Property slump 1991–1997 Property boom 1997–2000
2001 >	Globalization Sustained economic growth Low interest rates and inflation	e-Procurement Prime contracting Relationship contracting	Property boom

city housing; housing that had been built during the mid-19th century for industrial workers, flocking to the cities from the countryside, becomes available for compulsory purchase and redevelopment by local authorities. With an almost evangelical fervour most major cities in the UK were drawn into the spirit of the age and planned massive redevelopment programmes. With such a large programme of works the over-riding procurement strategy was lowest initial cost – so much had to be rebuilt and money was in short supply. Considerations such as life cycle costs were very much in their infancy and unfortunately the legacy of this time can still be seen in most major UK cities in the form of flatted tower blocks with major maintenance and running cost problems. The rebuilding programme was given further impetus with the election of a Labour government in 1964 which, at the same time, sought to shackle developers from making what were considered to be excessive profits, with the introduction of a Betterment Levy, that was included in the Land Commission Act of 1967. The effect of the land commission was to stifle development activity in the private sector and dampened prices until its dissolution in 1971, when the genie was let of out the bottle! This period, immediately prior to the development of information technology was the golden age of the bill of quantities. Even modest projects routinely had lead-in times of several months or even years, while each post and posthole was drawn, specified, counted and recorded in bills of quantities. Despite this period of sustained, continuous economic growth, several government sponsored reports, namely Simon (1944), Emerson (1962) and Banwell (1964) suggested that all was not well within the construction industry and pointed the finger particularly at procurement. But the reports were without teeth and were in the main confined to the filing cabinet. This period was also a golden age for construction professionals, operating on generous fee scales, before compulsory competitive tendering was to savage income flow. The predominant forms of procurement/contract during this period were single stage lump sum contracts based on bills of quantities. A Code of Practice for Single Stage Selective Tendering was developed as a highly prescriptive guide to this form of procurement.

This period saw the emergence of cost reimbursement contracts which allowed a contractor to be reimbursed for the costs of a project on the basis of actual cost of labour, material and plant plus a previously agreed percentage, to cover profit and overheads. The theory was then work could commence on site much more quickly than traditional procurement routes, although the higher the cost the greater the contractor's profit. Cost reimbursement contracts were only used *'where a special relationship exists between the employer and the building contractor'* (Ashworth), or in other words, when the contractor could be trusted, which perhaps says a lot about construction industry relationships generally.

1970–1979: a roller coaster ride

1970 saw the election of a Conservative government and the start of a property boom. As previously discussed, the 1960s had seen a Labour government artificially restrain property development in a corset of legislation as well as controlling house building with mortgage lending restrictions. What took place during 1970–1974 was a classic bust to boom to bust property cycle, a cycle that was to repeat itself almost 20 years later. The Conservative government decided to increase the money supply and much of this money found its way into property development. Between 1970 and 1973 bank lending increased from £71 to £1332 billion, with most of the increase going to property companies. There was no lack of demand and institutional investors were willing to buy all completed developments. Consequently rents doubled within 2 years. It was during this period that developers became increasingly impatient with existing procurement techniques, they demanded short lead-in times and fast completion. This pressure was given further impetus by rampant inflation of up to 25% per annum and historically high interest rates. The faster that buildings could be procured the less profits were eaten up by interest charges and inflated material and labour prices. During this period it was common for quantity surveyors to allow 2% per month for inflation on all approximate estimates and cost plans – no wonder that developers wanted to get on with it! The markets were optimistic that increasing rental and property values would offset rising prices, but in mid-1972 the continuing Arab/Israeli war caused oil prices to soar and a loss of confidence in the UK economy. Against this there was a period of comparative political instability with the election of a minority Labour government in February 1974 followed by another

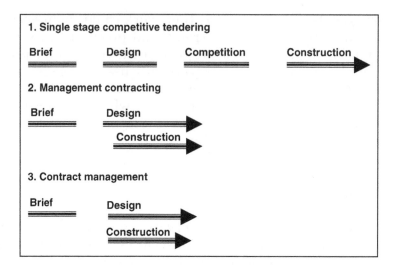

Figure 1.5 Procurement strategies compared.

general election in October 1974. Labour was to remain in power until 1979 when a new Conservative government was elected. Predominant forms of procurement/contract during this period were, in the private sector, fast track procurement strategies. Although faster than traditional methods they had little cost certainty and were a 'Pandora's Box' of unknowns, particularly for refurbishment projects. However, lack of cost certainty was to some extent tolerated in times when property values were rising so quickly. Many of the problems that were to be associated with the so-called fast track methods of construction management and management contracting were due to lack of precontract planning. The traditional approach to procurement is based on sequential operations; these new methods, however, sought to collapse the construction programme by using a concurrent approach with in some cases design and construction being carried on in parallel (see Figure 1.5).

In the public sector traditional procurement, that is single stage selective tendering, still predominated in an attempt to ensure accountability.

1980–89: a difficult time

This period started with high interest rates that contributed to the 1979–1982 recession, however just as had happened in the

1970s banks, including foreign banks, transformed themselves from being conservative risk managers to target-driven loan sellers.

The emergence of the developer-trader. Often an individual, who had as many projects as money could be found for, the completed projects were sold for a profit at a time of rising rental values, the bank repaid and the next project undertaken. This client demanded fast track methods and was not at all interested in prolonged precontract development, bills of quantities, consideration of life cycle costs, uncertainty in completion, countless remeasurements and fully documented final accounts. By the end of the decade UK banks were exposed to £500 billion of property related debt. It was noted previously that the immediate post-war period was a golden time for consultants with most work being awarded on generous fee scales. During the 1980s the Conservative government introduced Compulsory Competitive Tendering (CCT) for public sector projects which, by the end of the decade, spread to the private sector and consultants, faced with a reduction of fee income of up to 60% had to adopt a more pragmatic approach to procurement.

The later part of the 1980s was characterized by the increasing use of modified standard forms of contract. Private clients/developers became increasingly frustrated with what was perceived as unfair risk allocation and as a result lawyers were increasingly asked to amend, in some cases substantially, standard forms of contract to redress the balance. The chickens came home to roost, however, years later when these amendments were tested in the courts. Also during this period Design and Build procurement began to see a rise in popularity as clients perceived it as a strategy that transfers more risk to the contractor.

1990–present: the re-modelling of the UK construction industry

The construction industry was not the only sector experiencing difficult trading conditions. In 1992 the oil and gas industries came to the conclusion that the gravy train had finally hit the buffers. It launched an initiative known as Cost Reduction Initiative for the New Era (CRINE), with the stated objective of reducing development and production costs by 30%. The chief weapon in this initiative was partnering and collaborative working in place of silo working and confrontation and the pooling of information for the common good. It worked. In 1999 the construction industry started to come to terms with the new order. A new breed of client was in the ascendancy, demanding value for money and consideration of factors such as life cycle costs. Predominant forms of procurement/contract during this period were partnering, PPP/PFI, and Design and Build, although traditional lump sum procurement still accounted for approximately 48%, in terms of value, of all procurement in 2002.

Civil engineering procurement

Construction is not just building; civil engineering is a very important part of total construction activity in the UK. Another interesting perspective of recent changes in procurement trends comes from the *Civil Engineering Industry in the Civil Engineering Contractors Association (CECA) Report – Supply Chain Relationships (October 2002)*. The annual output of the civil engineering contracting industry is estimated by the CECA to be between 18% and 19% of total construction output. With sub-contracted work included, the civil engineering industry's output is somewhat higher at around £21 billion in 2001/2002. The civil engineering industry importance lies in the fact that the demand for its services comes from a relatively small number of large high profile clients including Railtrack, The Highways Agency and British Airports Authority. It is largely these large repeat customers that have been forcing the pace in terms of the introduction of new procurement strategies and new forms of contract and according to the CECA have led to a crisis of confidence within large parts of the UK civil engineering industry. Again according to the CECA the civil engineering industry generally has no problem with the main objectives of value for money and continuous improvement, where there are problems however, it is with the extent to which clients fully understand the changes they are introducing and are capable of managing them and the extent to which contractors, accustomed to traditional, adversarial relationships are willing and able to accept and adapt to the new approach. Sadly it is the conclusion of the CECA that, nearly 9 years on from Latham, it is lack of trust between clients, main

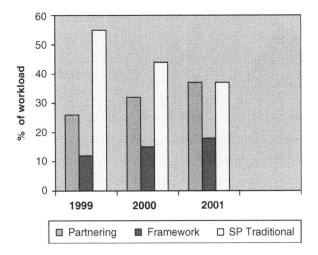

Figure 1.6 CECA survey of contractors.

contractors and supply chain partners that remains as the main stumbling block to improvement in procurement performance. Nevertheless there is evidence that, in the civil engineering industry at least, forms of procurement are shifting to a less adversarial approach. Figure 1.6 illustrates the findings of a survey carried out by the CECA in 2002 among larger civil engineering contractors, that is contractors with a turnover of more than £25 million per annum.

Among the larger contractors that took part in the survey:

- 56% reported an increase between 1999 and 2001 in the proportion of their workload carried out under partnering arrangements; see Chapter 5, from 26% to 37%;
- 52% reported an increase in the proportion of their work carried out under framework agreements, with the average rising from 12% to 18%;
- 74% reported a fall in the proportion carried out under traditional contracts with the average proportion down from 55% to 37%.

Best value

Another area of uncertainty in approaches to civil engineering procurement has been created by the introduction of the best

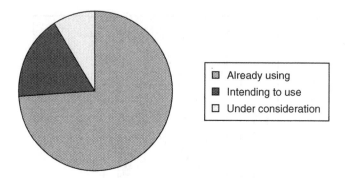

Figure 1.7 Local authorities' use of partnering (source: CECA).

value regime for the delivery of services by local authorities and certain other public bodies in England and Wales, introduced under the Local Government Act 1999, which came into force on 1 April 2000. This replaced the CCT requirements for the award of works and services contracts by local authorities that formed a major plank in the previous Conservative government's local government policy platform. According to the CECA two-and-a-half years after its introduction there are still uncertainties over how authorities will implement best value particularly in respect of the procurement of services such as highway maintenance. Despite this, there is strong evidence from the CECA that there is already a strong trend in favour of partnering among local authorities as illustrated in Figure 1.7.

Lean procurement

The construction industry is a service provider and in common with other industries is concerned with providing value. However, because of the diverse nature of the products there is no clear focus or definition as to what produces value, or more significantly, how to go about it.

Winch *et al.* (2003) suggest that there are four aspects to buildings that add value for clients, namely (Figure 1.8):

- Spatial quality, measured in terms of the spatial configuration of the facility and its urban environment designed to encourage interaction between staff or to reduce crime, etc.

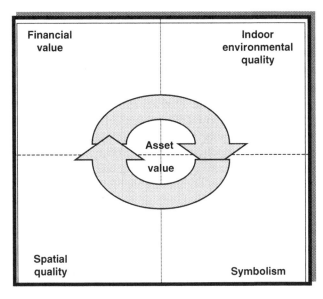

Figure 1.8 Adding value in buildings (source: Spencer and Winch, 2002).

- Indoor environmental quality, and its impact on the efficiency and the effectiveness of the people who work in the facility.
- Symbolism in terms of the extent to which the facility communicates the identity and the values of its owners.
- Financial value, a capital asset for exploitation or sale.

Could this model replace the time, cost, quality model referred to earlier, with clients being asked which of the four value producing aspects is paramount?

Perhaps one of the more controversial suggestions to be laid at the table of UK construction during the current value for money debate is that construction should adopt production and manufacturing approaches to procurement, including lean production. The origins of this movement will now be briefly examined before the theory of lean production is discussed.

This book will not dwell on the much discussed reports, Latham (1994) or Egan – Rethinking Construction (1998), beyond recommending them as background reading to the value for money in construction debate. In 1997 Sir John Egan was appointed as chairman of the Construction Task Force. Egan's approach to construction industry reform, clearly based upon his years as Chairman and Managing Director of Jaguar in the 1980s, was signalled, when he visited the Nissan motor car plant as one of his first public duties. His visit was to highlight the ways in which supply chain management and lean thinking had been applied in motor car production with apparent success. Subsequently, the Strategic Forum for Construction was established and the Rethinking Construction Report was published (1998) and is significant, because it shaped the agenda for change in the UK construction industry and was heavily supported and promoted by the UK government and various influential agencies.

Despite the evidence there remains a substantial body of opinion that doubts whether lean production methods are truly applicable outside Japan. For example, it is widely accepted that just in time (JIT) inventory management, one of the main tools in the lean toolbox, that originated in the Japanese shipyards in the 1950s could not have prospered in any other culture than Japan. Indeed in the preface to *The Machine that Changed the World* Womack, J.P. *et al.*, the authors thought it necessary to insert a paragraph asking readers not to dismiss the work as just another 'Japan' book, that is to say '*concerned with how a sub-set of the population within a relatively small country manufactured goods in a unique way.*' In a similar vein Green (1999) warned that the lean methods may not be transferable beyond the specific context of Japanese motor manufacture and that Egan zealots have ignored the extensive debate regarding lean transferability. There have also been concerns expressed (Green, 1999) about the wider social, moral and political contexts of lean production techniques and its effect on the workforce of other industrial sectors that have adopted the lean approach.

Early usage of the lean approach decentralized management and involved a movement towards transparency. The task of engineering components to meet design and production criteria was shifted to the suppliers. New commercial/partnering contracts were introduced which gave the suppliers incentive continually to reduce both the cost of their components and to participate in the overall improvement of the product and the delivery process. Partnering makes great sense from an activity perspective, however, few realize partnering is a solution to the failure of central control to manage production in conditions of high uncertainty and complexity.

Despite the reservations concerning the lean approach, Rethinking Construction received widespread acceptance from influential sections of the UK construction industry as well as government and was the basis of a number of government based initiatives that followed during the early part of the new millennium. Subsequently, during Egan's period as chairman of the Construction Task Force, 'Lean Thinking' became a generic term to symbolize the drive towards a more efficient, less wasteful, leaner construction industry. However, as Egan stepped aside for the new Chairman of the Strategic Forum, Peter Rogers, in September 2002, concerns were already being expressed about some aspects of the lean approach and its applicability to construction. Egan's departure was at a time when it was thought by many within the industry that the agenda for change was losing momentum and the title of the forum's last report under Egan's pilotage 'Accelerating Change' seemed to some observers to articulate Egan's frustration. Accelerating Change was a report in two parts, the first part, a consultation document was published in April 2002 to quote Egan at *'a real moment of change for the UK construction industry.'* Responses to this document were sort and subsequently in September 2002 Accelerating Change was published. The report restated the need for the UK construction industry to change its culture in order to become world class. Chapter 2 of Accelerating Change contains a section entitled 'Progress since Rethinking Construction'. This is largely concerned with the results from the M4i's demonstration projects programme, which implements one of the key Rethinking Construction recommendations. While the results of the demonstration projects are clearly encouraging, 400 projects and £6 billion worth of projects still covers only a very small proportion of total construction activity. According to the report the principal ways in which improvement can be achieved are identified as follows:

- Greater client focus and client leadership and an end to lowest cost tendering – described as wasteful and unpredictable – to be replaced with clients procuring value for money.
- Continuous improvement measured against world class benchmarks, see Chapter 6.
- Consideration of whole life costs, see Chapter 3.
- Improved health and safety record, which was suggested would result from better planned and organized projects.

The route to achieving these goals is being heavily geared to the adoption of lean construction techniques. Accelerating Change recognized the diversity of construction clients and that the larger clients with in-house teams will almost certainly be able to establish the critical development information necessary to deliver value for money. For the occasional client however according to the report, this advice would be hard to find and to fill this gap a new role of 'Independent Client Advisor' was proposed by Egan. The mission of this new design team member was undefined, but appeared to focus on procurement and design issues. In addition getting this message across to an industry, where so many of the enterprises are small, (see earlier in the chapter) requires a special type of approach, which was also undefined, although a promise was given to 'educate' construction enterprises on how to introduce supply chain management. However, the key measure identified by the Accelerating Change Report is the adoption and extension of collaborative working within the industry. The case for adopting collaborative working is built primarily around the previously mentioned pilot or demonstration projects. The pilot project's results were interpolated against the performance of the UK construction industry as a whole resulting, it was suggested, in a possible substantial increase in profits and reduced costs as well as accidents. Just as a decade earlier Latham had set out performance targets in constructing the team, so did Accelerating Change:

- By the end of 2004, 20% of construction projects by value should be undertaken as integrated teams and supply chains. The target is to rise to 50% by 2007.
- By the end of 2003 project insurance should be used on all demonstration projects – this will be discussed in full later in Chapter 2.

The publication of Accelerating Change and the long awaited guidance on modernizing construction supply chains was branded by many as 'hopelessly optimistic' (Supply Management, 19.09.02). There has also been strong criticism from other sources. Green (1999) declared that *'the whole concept of lean thinking and the case for its application to the construction industry has not been proved.'* The current agenda, Green argues, ignores the extensive critical literature on lean production and in addition it is contended, relies on a pattern of a regressive approach to human resource management developed in Japan.

Evolution of Keiretsu and their different forms

As has been previously mentioned, several commentators on the evolution of Japanese business practices have drawn attention to the unique pattern of social and economic influences that occur in that country. Some of these will now be briefly discussed as background to the debate. One of the most influential features of business culture in Japan are the Keiretsu.

Japan before the war was dominated by four large Zaibatsus: Mitsubishi, Mitsui, Sumitomo and Yasuda. These were involved in steel, international trading, banking and other key sectors in the economy and controlled by a holding company which established financial links between the different members. Large, influential banks were part of these conglomerates, providing necessary funds. At the end of the war the occupational forces decided that these structures had to be broken up, since they constituted powerful monopolies that helped the former government to execute the war. As a result it was planned to sell shares to the public, to remove executives from their positions and so forth. In 1947 a law was enacted to dissolve large companies and enhance competition. However, in 1948 the allied forces realized that they needed a strong Japan to fight the Korean war and communism in general. Therefore they stopped weakening the Japanese economy and conversely pumped substantial aid into it. Few companies were broken up and in the amendments of anti-trust laws in the aftermath, many of them were re-established. This time, companies grouped round the banks that were then allowed to hold shares in other companies, which made the establishment of financial links easier. These conglomerates were now called Keiretsu. Some emerged out of former Zaibatsus but others were just new groupings of companies. Different forms emerged, where the most important ones are the 'financial Keiretsus' (Big Six) with horizontal relationships across industries. The ex-Zaibatsu – Mitsui, Mitsubishi and Sumitomo – belong to this grouping with close-knit relationships among group members. Sanwa, Fuyo and Ikkan are newly formed groups with a bank at their core. Usually they only have one enterprise in each business sector to enjoy economies of scale and avoid competition within the group. Another form is represented by the 'distribution' Keiretsu with vertical relationships, controlling the way the flow of products, services, prices, etc. from the factory to the consumer. In most cases they are smaller than the horizontal groupings or even part of them. One large company is a member of a horizontal Keiretsu but has its own independent vertical group. Generally distribution Keiretsus are less influenced by a bank. The advantage of this structure lays on the grounds of many sub-contractors working as 'buffers' for the core firm in economic downturns. They do so because they can fire their temporary workers at any time as opposed to the lifetime employment situation in the large companies. As a result the whole group can adapt to changes in the business environment. Part of the formation, symbolized by vertical relationships, are the 'manufacturers' Keiretsus integrating a pyramid of suppliers and component manufacturers in one structure. They behave as if they were one company: giving loans, technology, development costs, long-term supply agreements, etc. from customers higher up in the pyramid to sub-contractors. The latter even absorb losses occurring in other sectors and pursue set prices. The result is a conformist structure producing high quality while shutting out foreign suppliers. Many other forms of cartels and groupings are common in Japan. The Japanese believe that they ensure full employment, the security of the nation and distribute risks; it is their version of capitalism.

The nature of the Keiretsu

At the core of the 'Big Six' Keiretsu are a bank and a trading company. Japanese banks are permitted to have equity in other firms, although this is limited to less than 5% of the total number of shares issued by a company. The shares of group affiliated members that belong to the bank create an important link in the close, interlocking nature of the Keiretsu. Banks also appraise the investment projects planned by group firms and provide preferential loans when necessary. Furthermore, low profit and poor dividends will be tolerated by these financial institutions, in so far as they are a result of costs relating to long-term projects. The general trading companies focus on the import and export of a wide range of commodities throughout the world, utilizing their large network of contacts to acquire information relating to their business. Each of the 'Big Six' Keiretsu groups also possesses a President's Club, which provides a forum for core members to meet in order to determine the strategic direction of the group. A major characteristic of a Keiretsu is the cross holding of shares (*interlocking shares*). Group affiliated firms issue shares and assign them to member firms, a measure that was initially established by Article 280 of the Commerce Law in order to prevent takeover threats from foreign corporations. By placing large amounts of stock in the hands of these 'friendly' shareholders who will be reluctant to trade the shares, it is not only difficult for foreign companies to acquire the level of shares required to take over a company, but also management is not pressured to achieve short-term profit at the expense of long-term growth. In addition, interlocking shares act as a vehicle for the close monitoring and potential disciplining of group affiliated firms. The interlocking shares ratio (the ratio of shares owned by other group firms to total shares issued) and the intragroup loan ratio (the ratio of loans received from financial institutions in the group to total loans received) are often utilized as a measure of the strength of the relationship between member companies. They are a strong indication of the level of group orientation. The large horizontal Keiretsu span a wide range of industries, including banking, insurance, steel, trading, manufacturing, electric, gas and chemicals. Where possible, groups try to avoid direct competition between member firms by having only one company in any category of business. This is known as the 'one-set principle' and is adhered to as much as is possible. The large Keiretsu have an enormous influence on Japanese industrial and economic policy. The roots of their influence stems from the 1920s, when a very close relationship existed between government officials and their predecessors, the Zaibatsu. It is often the case that members will utilize the products or services of group members, rather than acquiring them from non-group members or foreign importers. The preferential buying habits of the Keiretsu have been cited by the Americans as a barrier to free trade, keeping foreign investors and foreign goods out of the Japanese market. Critics also argue that Keiretsu and the Japanese cartels help to keep Japanese prices high, since they are able firmly to control the price and distribution of products and services from the manufacturer to the consumer. It is believed that the existence of Keiretsu prevents American firms from undercutting these high prices. In fact, it is claimed that, if it were not for these 'exclusionary practices', Japan would have imported $40 billion more goods in 1987 (according to a report by the Brookings Institution). In these large groupings, there is a tendency for employees to remain in the same company for their entire working life. Redundancies are avoided through the transfer of workers from depressed to more prosperous industries within the Keiretsu network. This network also enables the cooperative development of new products and processes and the sharing of vital information, which encourages innovation, significantly reduces the costs of research and development and subsequently leads to higher quality products.

For example, Mitsubishi having been a Zaibatsu is now one of the 'Big Six' Keiretsus in Japan. It is the most tightly interwoven of Japan's corporate families and is a horizontal Keiretsu consisting of 29 companies and many subsidiaries with a high percentage of interlocking share ratios and intergroup loans. It has an empire of 216,000 employees, and is prominent in businesses ranging from banking to beer, shipping and shipbuilding, property, oil, aerospace and textiles in fact their old slogan was 'From noodle to atomic power'!!! Mitsubishi exchanges more directors than the other Keiretsu and have a long record of gathering companies from different parts of the group to cooperate on projects which is a main advantage of the Keiretsu system. Another quality of the system is the ability to pass information quickly around the corporate family. In the case of Mitsubishi there is a 'Friday Club' at which chief executives of the 29-member companies meet over lunch on the second Friday of each month on the top floor of the Mitsubishi Building in Marunouchi, this is also the home of their headquarters. It is here that the strategies for the group as a whole are decided. The close ties are continually reinforced, for example employees of the group do their personal banking with the Mitsubishi Bank. Indeed, in 1972 the 'Diamond Family Club' was created which promoted marriages between group members. This highlights an important difference between the culture of the West and Japan – their group spirit or sense of membership makes the power of mutual holding of stocks much stronger than it actually seems. One criticism of the Keiretsu is that because of the share interlocking they pay out low dividends, but on the other hand this is often compensated for by other commercial advantages, for example the availability of large sums for capital investment. The Landmark Tower development is a classic example of the power and resources that the Mitsubishi Keiretsu hold. The 70-storey tower in Yokohama will cost an estimated ¥270 billion. An independent company might have been forced to abandon the project when there was a 10% vacancy rate even during the property boom of the 1980s, yet Mitsubishi Estate was able to carry on, confident that Mitsubishi Bank would continue funding the project through good times and bad. Indeed, 'Even the most critical property analysts believe that any company that can afford to continue owning the tower will end up, in a few years, with one of the finest assets in the domestic property market'. It is then easy to understand the reasons why SCM prospered in Japan, especially when combined with the pluralistic approach to decision making fostered by Japanese management schools.

What is lean?

To many observers the definition of lean construction remains illusive. The Egan Report however seemed to have no such doubts:

> We are impressed by the dramatic success being achieved by leading companies that are implementing the principles of 'lean thinking' and we believe that the concept holds much promise for construction.

According to Ballard and Howell (2003) the phrase 'lean production' was coined by a member of the research team studying the international automobile industry: the report of which was published in *The Machine that Changed the World* (Womack *et al.*, 1990, 1991).

Lean production can be described as a new way to design and make things. Following the publication by Lauri Koskela in 1992 of *The Application of the New Production Philosophy to Construction*, a number of organizations worldwide were established to promote the lean philosophy in construction. The most notable of which are The International Group for Lean

Construction and the Lean Construction Institute. These organizations draw heavily on the reported success of lean production in the car industry and the work done by Womack *et al.* (1990) at the Massachusetts Institute of Technology in the 1980s. Since the mid-1990s lean construction has emerged as a new concept. According to Koskela *et al.* (2002) there are two slightly different interpretations of lean construction. The first is that lean is concerned with the application of lean production methods to construction, the second regards lean as the move to a new theory based methodology for construction, as expounded by Lauri Koskela. Lean construction is still to a certain extent, 'work in progress' and a developing area. Production management is at the heart of lean construction and runs from the very beginning of a project to the handover of the facility to the client. By contrast Green (1999), observes lean thinking comprises a complex cocktail of ideas and thinking including: continuous improvement, flattened organizational structures, the elimination of waste, efficient use of resources and cooperative supply chain management that became synonymous with best practice in construction industry circles in the late 1990s. According to Howell (1999) lean supports the development of teamwork and a willingness to shift burdens along the supply chains. It is claimed that partnering relationships coupled with lean thinking make rapid implementation possible. Figure 1.9 illustrates the principles of lean construction. The five triads represent the four stages in project delivery, for example predesign, design, procurement and delivery, plus a fifth for operations, maintenance, etc. The third, procurement triad, can be seen to incorporate the simultaneous operations of:

- product design,
- fabrication and delivery,
- detailed engineering,

with, it is assumed, is a preselected supply chain cluster. It is claimed that traditional procurement practice requires the design to be frozen at some point in order that the bills of quantities and bids may be obtained. Whereas lean approaches enable and encourage continuous development and design teams/supply chains are actively encouraged to leave decisions and choices until the last possible moment. In this respect lean is very similar to traditional supply chain management.

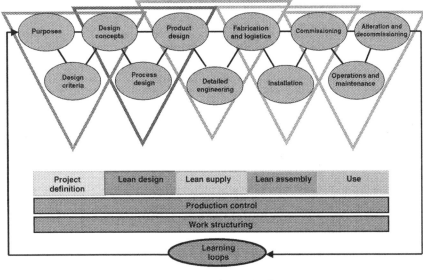

Figure 1.9 Triads of lean construction (source: Lean Construction Institute, with kind permission of Greg Howell).

Womack *et al.* (1990) summarized the benefits of lean production as follows:

- It transfers the maximum number of tasks and responsibilities to those workers (suppliers) responsible to those workers actually adding value to the product.
- It has in place a system for detecting defects that quickly traces every problem, once discovered, to its ultimate cause.
- This in turn results in teamwork among the suppliers that make it possible for everyone in the supply chain to respond quickly to problems and to understand the project's overall situation.
- Rather than members of the supply chain jealously guarding information on the grounds of commercial advantage, knowledge is shared among the supply chain members.
- Problems associated with defects or problems with different parts being incompatible were originally highlighted by senior management, who were the only people charged with the authority to spot the potential problem. Toyota, example, gave the production line worker the power to stop the line and discuss any problems – however in practice the

Table 1.4 Product development performance

	Japan	USA	Europe
Average engineering hours per new car (millions)	1.7	3.1	2.9
Average development time (months)	46.2	60.4	57.3
Number of employees in project team	485	903	904
Time from production start to first sale (months)	1	4	2

Source: Womack *et al.* (1990, 1991).

production line seldom stops as problems are solved in advance. The result:

- Every member of the supply chain understands the operational circumstances of the other members allowing simultaneous development of design and components.
- How has lean production impacted on other industries? (see Table 1.4)

The five principles of lean production techniques have been articulated by Womack and Jones (1996) as follows:

- *Greater customer focus when specifying value* Value being defined by the ultimate customer's needs through tools such as value management and simulation (modelling techniques) and comes before design.
- *Identify and map the value stream* For example, by examining elements such as the structure and the building envelope and considering how these systems are to be designed, supplied and constructed. It is said to identify how value is created and built into the product from the client's perspective and identifies choices.
- *Flows* Strategically, flow is concerned with achieving a holistic route through the means by which a product is developed. It attempts to reduce waste and fragmentation and latterly in the construction industry the adoption of partnering has gone some way in achieving this. It is claimed that improvement can be achieved by reducing uncertainties in workflow and moving away from activity based management systems, such as critical path analysis. It is said that lean aims to have the labour force working

continuously and evenly, without having to wait for materials or upstream activities to be completed. One way it is suggested that this could be implemented is by maintaining excess capacity in the labour force so that it can speed up or slow down as conditions dictate. In stable circumstances managers can predict the work content and shift suppliers to minimize imbalance, however a factory environment has little to do with construction!

- *Pull* The process by which products are delivered to the customer. It focuses on the elimination of waste in the supply chain by reducing unnecessary production, transporting and storing of goods.
- *Perfection* Seen as the ultimate goal of the construction process – to produce a built asset with as few defects as a Toyota car. The means to this end are ISO certification, benchmarking and continuous improvement and ultimately a new construction culture.

Is lean in danger of discrediting the drive towards a more efficient, more client focused industry?

The case for alignment of construction procurement to manufacturing

As with all industries a high level of productivity in the construction industry is a very important driver in maintaining its global position and competitiveness. In construction this manifests itself in the procurement and delivery of projects to time, to budget and with implementation of sustainability. The post-Latham era saw so many initiatives that the construction industry could not see the wood for the trees and initiatives like lean construction and the setting of overly optimistic targets were in danger of being seized by the Luddites as an excuse to maintain the *status quo*.

The first goal of lean construction is to understand the so-called 'physics' of production and the effects of dependence and variation along supply and assembly chains (Howell, 1999). It is argued that with current approaches the physics of production are ignored, instead current practice tends to focus on teamwork, communication and commercial contracts. Lean construction is generally accepted to mean, construction with minimum waste and has the goal of producing built assets that better meet customer needs over its life cycle. It is, according to G.A. Howell, Director of the Lean Construction Institute, *'most suited to complex, uncertain and quick projects.'* In the context of lean, waste is defined as failure to meet the unique requirements of a client. Another characteristic of lean is that focus is shifted away from the activity to the delivery system.

The proponents of lean construction claim that the traditional approach to construction procurement is as a number of discrete steps of which procurement is one, each independently adding to the value of the product and is based on the pull model, for example:

- Break down project into tasks, for example Superstructure, Finishes, etc.
- Break down into smaller tasks.
- Allocate resources to each task.

Ignoring both value maximization and waste minimization: Optimizing each or any of the operations will move the process as a whole towards an optimized condition. Lowest price for each operation, order, contract or purchase is expected to lead inevitably to the lowest total cost for the project as a whole. However, analysis has shown that there are likely to be many more non-value adding activities (waste) than value adding ones and that the UK construction industry has still to understand this. Lean construction (Koskela, 1992) requires the construction process to be viewed as a flow and conversion process, instead of current thinking that has its focus on construction as an activity based process and assumes that customer value has been identified during the design process, which of course is often far from the truth. Subsequently, for example; production is managed throughout a project by first dividing the project into packages based usually on trades, arranging these packages in a logical sequence together with an estimate of the time and resources required to complete each activity and therefore the project. Each package is let to sub-contractors or specialists. Control is perceived as monitoring each package against its schedule and cost targets. If packages fall behind the programme or critical path, or appear to be exceeding cost targets, efforts are made to reduce costs or speeding up the offending package – usually at an extra cost.

This concentration on activities conceals the waste generated between continuing activities by the unpredictable release of work, that is, downstream suppliers waiting for upstream suppliers to complete an activity and/or the delivery of materials. Proponents of lean construction see the problem as one of dependency on other activities. An essential part of lean thinking therefore, is the management of the interaction between activities and the combined effects of the dependence and variation. Minimizing the combined effects of the dependence and variation becomes a central issue for the planning and control system as project duration is reduced and the complexity increases. Complexity in this context being defined by the number of activities that can interact. Ten years after its first introduction it is significant that papers still have to be produced explaining the concept of lean construction.

According to Bertelsen and Koskela (2002) the work on lean construction has to date to a great extent been focused on two major areas the understanding of the application and implementation of the new production principles in construction, which taken together have become known as lean project delivery system (LPDS). This LPDS approach is based on the following:

The first area is *work structuring – push model*:

* Understanding construction as a production and planning the Transformation, Flow and Value (TFV) concept approach – Koskela *et al.* (1992). According to Bertelsen and Koskela (2002) there are three different conceptualizations of production that are used in practice. The first (T) is viewing production as the transfer of inputs to outputs. The second conceptualization views production as a flow (F) where in addition to transformations there are waiting, inspection and moving stages. The third conceptualization views production as a means for the fulfilment of the customer's needs. It is argued that these three conceptualizations should be used simultaneously resulting in TFV (see Figure 1.10).

The second main area of lean construction is *production control – the Last Planner System:*

* A system of lean working project management that produces a range of outputs including:
  – project execution strategies,

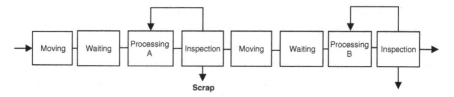

Figure 1.10 Production flow as understood by Koskela (2000) (source: Bertelsen, 2002).

- project organizational structures, including configuration of supply chains,
- operations designs,
- master schedules,
- phase schedules,

that taken together form a temporary production system.

• Managing the workflow within the construction process – the so-called last planner (LP) approach (Ballard *et al.*). The LP system guide devised by Greg Howell and Hal Macomber of Lean Projects Consulting Inc. explains this approach as follows.

The general idea behind the LP System is that in order to obtain an even workflow weekly work planning and monitoring of planned performance is needed. The LP identifies what will be done and attempts to ensure even workflow. With LP a construction project is likened to a set of promises – the bigger the project, the bigger the promises. It also appears to rely heavily on the ethos of trust and partnering as well as organizing the progress of a construction project using a pull schedule instead of the traditional push approach. A complete pull schedule represents the design of the production system in terms of the work done by each trade or specialist and establishes the conditions for release of work. Each party should understand and support the schedule, that is, both how their effort contributes to the larger picture and the nature of their commitment to the project and downstream workers. A pull schedule is a promise from each supplier that they can complete the work in the time allocated – it is prepared by starting from the end milestones and working backwards. It is dependent on all the suppliers being consulted during this process. Central to successful implementation of LP are a series of systems of constant review known as 'Look-Ahead

Planning' which support the practice of making work ready and controlling the flow of work through the production system and 'Weekly Work Planning' which can be thought of as assessing what promises each supplier is making for the coming week. Several projects worldwide are using lean approaches to procurement and delivery apparently with impressive results.

Experience from industry

Lean thinking and supply chain management do have an important role in delivering client value, for one essential reason – changes in the method and type of procurement are only likely to occur as a result of enhanced profitability – supply chain management can do this. All parties do not accept the argument that the construction process is an assembly industry and not a manufacturing industry, in that it assembles components that are manufactured by a variety of suppliers. Manufacturing is usually concerned with the production of a range of standard products or models and investment in the form of plant and machinery lends itself more easily to manufacturing industries and less to the assembly process. Supply chain management focuses the supply chain on delivering value.

In 2000, Common *et al.* conducted a survey of 50 large UK companies to discover the penetration of lean thinking. The survey and resulting analysis revealed a distinct lack of understanding and application of the fundamental techniques required for a lean culture to exist and that there appears to be significantly less lean culture in UK construction companies than is professed.

Is Egan too impatient? Change takes a long time according to two government agencies, the Highways and Environment Agencies that have tried to adopt new procurement strategies. The Highways and Environment Agencies are two government agencies carrying out major and diverse projects ranging from £10,000–100 million who recognized that a change in approach to the procurement of built assets was long overdue.

They found that their approach to procurement was highly traditional and in their experience of construction produced:

- Little job satisfaction
- Single project focus
- High claims exposure (£40 million plus) per annum – Environment Agency

- Little understanding by the supply chain of underlying environmental issues – Environment Agency
- A focus on cost not value
- Poor buildability
- Traditional attitude to risk.

In 2002 new procurement strategies were introduced by both organizations based upon:

- Best value
- Alignment of client and supply chain objectives together with continuous measurement and improvement
- Better planning
- Long-term relationships.

The key success factors for the effective implementation of new procurement strategies have been identified by the agencies as:

- Training/communication
- Open book accounting
- Bench marking
  Incentivation at various levels
  – Framework level – profit level
  – Individual level – pain and gain.

Conclusions

The main challenges in procurement of built assets are overcoming the inertia and convincing the vast majority of UK construction enterprises that improvement is necessary.

Of all the current solutions that are being proffered to improve the efficiency of procurement in the UK construction industry perhaps the system of lean thinking is the one that is the most difficult to apply. It would have been better if this term, so long linked to the Japanese manufacturing sector, had not been seized on so readily by Sir John Egan to become a symbol for the many government-driven initiatives. The danger is that over-complexity of lean concepts may hamper the adoption of related techniques that can make a genuine contribution to procurement practice; techniques such as supply chain management and collaborative working. There is a danger that these approaches may be dismissed by a traditionally conservative construction industry as radical nonsense with no commercial pay-off – because at the end of the day it is factors such as increased profitability and commercial advantage that will convert the sceptics. One possible approach is that supply chain management is decoupled from the lean bandwagon for the time being – the lean approach being a step too far for many. If lean is to remain as a focus for the construction sector then perhaps it could be defined in slightly less esoteric terms, for example *lean* is:

- Understanding value
- Measuring value
- Identifying the processes that deliver value
- Identifying those processes that do not add value – waste
- The elimination of waste
- Pursuing continuous improvement.

Once again using a motor car industry comparison, it is well known that for example Toyota's dominant equation relating to car manufacture is:

Profit = Selling price − Cost

As the selling price is more or less constant in the case of a motor car, Toyota concentrates its efforts on driving out waste, reducing the cost and thereby increasing profits. Some observers, as has already been noted in the chapter, regard this as not too difficult in an industry that produces a standard product with a comparatively short life span in a controlled environment.

By comparison the dominant equation in the construction industry is:

Tender (selling) price = Profit + Cost

Unlike a motor car, the cost of built assets are not generally fixed and are subjected, during the procurement and delivery stages, to the adverse effect of a number of externalities. In addition, in the traditionally under-managed supply chains, there has been little incentive for the contractor to reduce cost. The question seems to be, how can the cost of procuring built assets become more certain and predictable with a focus on driving out unwanted costs? Or, as some quarters would have us believe, is the construction industry so unique and unpredictable that these goals are unachievable?

The remainder of this book will endeavour to discover whether or not new approaches can be successfully adopted in the procurement of built assets and will analyse some of the new procurement strategies currently being used, many with great success.

Bibliography

Ashworth A. (1994). *Contractual Procedures in the Construction Industry*, 2nd Edition. Longman.

Auditor General (2001). *Modernising Construction*. HMSO.

Berger M. (1994). *Well Connected*. Japan Scope.

Bertelsen S. and Koskela L. (2002). Managing the three aspects of production in construction. *Proceedings of the 9th Annual Conference of the International Group for Lean Construction*.

Booker C. (2003). Extravagance in Edinburgh. *Sunday Telegraph* April 22, p. 24.

Building. *A Bit of Strategic Thinking*. 13.09.02. p. 3.

CECA (October 2002). *Supply Chain Relationships*. Civil Engineering Contractors Association.

Cutts, R. (1992). *Capitalism in Japan: Cartels and Keiretsu. Harvard Business Review*.

Dti (2001). *Construction Statistics Annual*. HMSO.

Flybjerg B. *et al.* (2003). *Megaprojects and Risk: An Anatomy of Ambition*. Cambridge University Press.

Green S.D. (1999). *The Future of Lean Construction: A Brave New World?*

Green S.D. (1999). The dark side of lean construction: exploitation and ideology. *Proceedings of the 7th International Lean Thinking Congress*.

Harvey J. (2000). *Urban Land Economics*, 5th Edition. Macmillan Press.

Howell G.A. (1999). What is lean construction? *Proceedings of the 7th International Conference for Lean Construction*, Berkeley, CA, 26–28 July.

Howell G.L. and Macomber H. (2002). *A Guide for New Users of the Last Planner System – Nine Steps for Success*. Lean Project Consulting Inc.

Inside the Charmed Circle, Japan's Industrial Structure (1991). *The Economist*, 5 January.

Koskela L. (1992). *The Application of the New Production Philosophy to Construction.* Technical Report No. 72, CIFE, Stanford University, CA.

Mitsubishi's Extended Family (1994). *The Financial Times* 30 November.

Richards M. (2003). Budget? Fudge it! *Building Magazine*, 9 May. pp. 49–53.

Simon P. *et al.* (1997). *Project Risk Analysis and Management.* The Association of Project Management.

The end of a dynasty. *Building Magazine*, 5.10.2001.

Whingeing, *Japanese–American Trade* (1991). *The Economist* 18 May.

Winch G. *et al.* (Ed.) (2003). *Building Research and Information*, Vol. 31, Spon Press.

Womack J.P. *et al.* (1990). *The Machine that Changed the World.* Harper Perennial.

Womack J.P. *et al.* (1991). *The Machine that Changed the World.* New York, 1st Harper Perennial Edition.

Web sites

www.clientsuccess.org.uk/
www.ensembl.org/
www.leanconstruction.org/

2

Procurement – risk

A risk-less approach would lead to nothing unusual
ever being built, which would be sad for mankind.
—Andrew Gay

Generally

There is nothing new about risk – investment decisions in the
private sector are generally characterized by risk. A capital
investment means committing funds to procuring an asset before
the actual outcome benefits are known for certain. A characteristic of any capital transaction is therefore that it involves risk to
some extent.

A widely accepted definition of risk is; an uncertain event or
set of circumstances, that should it occur, will have an effect on
the achievement of the project objectives.

The questions therefore that should be addressed are as
follows:

- What are the risks?
- What will their impact be?
- What is the likelihood of the risks occurring?

after which, risks should be transferred or allocated to the
most competent person or organization to manage them. The
probabilities of some events or risks occurring can in principle
be calculated, but in other cases the chances of an event occurring is inherently unknowable. This distinction is critical to
practical approaches to quantifying risk, but it is not so relevant to the question of, what risks should be considered

as candidates for transfer. In practice, investors may use quantitative techniques to assess risk where they consider it to be cost effective as a technique, but commercial judgement and political considerations will ultimately always be involved.

All construction risk cannot be lumped in together, as risk comes in all shapes and sizes. Therefore, for the purposes of this chapter, large scale public projects, or mega projects, which are defined as, projects such as the Channel Tunnel or the new Scottish Parliament and which, for a variety of reasons, most often make the headlines, have been separated from the relatively more modest private sector projects. All capital works projects whether in public or private sectors, mega project or modest project, involve inherent risks in the form of political or economic change, climate, technology, ground conditions, engineering uncertainties, errors, industrial disputes, land issues, environmental issues and many more. Arrow and Lind (1996) suggest that governments can better cope with risk than private investors and therefore government investments should not be evaluated by the same criterion used in private markets. More specifically, they argue that governments should ignore risk because as governments invest in a greater number of diverse projects they are able, therefore, to pool risks to a much greater extent than private investors. It is the range, diversity and in some case the unpredictability of construction related risk that some groups argue sets it apart from other industrial sectors. In order to achieve optimal outcomes the client must select the most appropriate procurement strategy for managing these risks. Increasingly, in addition, construction projects have to be delivered in an environment of uncertainty, driven by diverse stakeholder interests, shifting business or political imperatives and rapid technological change. The traditional risk transfer models used in construction have increasingly been shown to be inadequate to deal with these circumstances. Construction procurement embraces a wide range of strategies. For example, at one end of the spectrum are lump sum fixed price contracts, in which the contractor undertakes the whole of the work for a specific sum – the contractor carries nearly all the risks as in this form of procurement:

- The client pays the same regardless of how much work is needed.

- The contractor is motivated to obtain the most cost effective approaches.

At the other end of the spectrum are cost reimbursement/ cost plus contracts where the client carries nearly all the risks:

- The client pays for work done plus a sum to cover for the contractor's overheads and profit.
- The contractor may not be motivated to carry out work efficiently or cost effectively.

Large complex buildings have traditionally carried a high degree of risk. Once more consider the new Scottish Parliament Building at Holyrood, Edinburgh, referred to previously. Designed by Catalan architect Enric Miralles as *'the visual embodiment of exciting constitutional change.'* The project, initially estimated to cost between £10 and £40 million and scheduled to open in December 2002, is now rumoured to cost between £340 and £400 million and at the time that this book went to print was still not complete. Admittedly, the project has had more than its share of problems, with the untimely and tragic deaths of both Miralles and the Scottish First Minister, Donald Dewar; even so such a high profile story of extravagance and ineptitude is very bad news for the UK construction industry. It is only because the extra £300 million or so will come out of the Scottish block grant – in other words the UK taxpayer's pocket, that the work is able to continue but, presumably at the expense of alternative capital investment in Scotland such as; the building or refurbishing of schools, hospitals, roads, etc. Despite the sentiment expressed in some quarters that the excess cost of high profile public sector projects is largely forgotten if eventually the asset works effectively, the Scottish Parliament just goes to reinforce the image that for many, construction is still a very risky jump into the unknown and the UK construction industry is inadequate and unable to deliver high profile projects to time and on budget.

Little wonder then that clients in both public and private sectors are demanding procurement solutions that reduce risk and introduce more certainty. The private sector and shareholders in particular are however less forgiving than the man on the top of the Princess Street omnibus in Edinburgh, for they are seeking income and growth, ensuring in turn that companies have to focus on productivity, profit margins, return on capital, etc. by 'sweating the assets' – built or otherwise.

Public sector projects and risk

It is generally public sector projects that are most often in the spotlight when things go wrong and risks are underestimated. What appears to be particularly baffling is that the lessons from projects that go wrong are not learned so that the same mistakes are not replicated, but perhaps all is not as it seems! Controversially, Flyvbjerg *et al.* conclude that in the case of public sector mega projects it is not the risk management that is poor, but over-runs in cost and time are to a large extent the result of promoters 'cooking the figures' in the belief that the project, when complete, will be of benefit to society. The same sort of philosophy and ambition that has driven the Presidential Project schemes in France and in particular Paris, during the past few decades, for example, Le Centre George Pompidou and La Grande Arche de La Défense. Flyvbjerg concludes that, as a result of analysing over 240 mega projects world wide, that the poor performance of these projects is not due solely to inadequacies of the risk management but the *'the dubious and widespread practices for underestimating costs and overestimating benefits used by the project promoters and forecasters to promote their favourite project, create a distorted hall of mirrors in which it is exceedingly difficult to decide which projects deserve undertaking and which not.'*

In a similar, but somewhat less outspoken vein, Mott MacDonald was commissioned to carry out a study for HM Treasury into the outcome of large public sector procurement in the UK during the past 20 years. The project sample included 80 projects, evenly spread across government departments with values exceeding £40 million, based on 2001 prices. It included private finance initiative (PFI) projects although it did not include what could be defined as a truly mega project in the terms of the Flyvbjerg study. Also, unlike the Flyvbjerg study the report did not concentrate on the possible political motives for the unsatisfactory outcome of so many high profile projects, but more the approach by civil servants to asses risk and the feasibility of such projects and the techniques that were used. The report did not detect any willful deceit on behalf of project sponsors although it did note that: '*Once a project has gained momentum (especially political), it is sometimes difficult to consider an alternative and so ultimately, the project goes ahead despite knowingly underestimating project costs and time*' – see cost benefit analysis (CBA) and the 2012 Olympic bid, later. The Mott MacDonald report highlighted more, a lack of skill and awareness on behalf of those concerned with the planning and development of large scale public projects and of the effects of their optimism, (for optimism read naivety) when appraising the project. The report continued by identifying the critical project risk areas most likely to cause over-runs of time and cost if sufficient risk mitigation strategies are not in place:

1. Inadequacy of the business case – 58%
2. Environmental impact – 19%
3. Disputes and claims – 16%
4. Economics (macro economic business cycle) – 13%
5. Late contractors' involvement in design – 12%
6. Complexity of contract structure – 11%
7. Legislation – 7%
8. Degree of innovation – 7%
9. Poor contractor capabilities – 6%
10. Project management team – 4%
11. Poor project intelligence – 4%
12. Other project risk areas – 3%

The poor performance of large public sector projects was euphemistically given the title of 'optimism bias' by Mott MacDonald which it was found occurred at various levels according to the project type, as illustrated in Table 2.1. Note that the lower the percentage the better the outcome. Even allowing for the staggeringly high optimism bias of 214% for equipment capital expenditure of traditionally procured projects, it would appear that the case for the introduction of the more rigorous approach adopted in public–private partnership PPP/PFI projects does produce more predictable results.

The Mott MacDonald report concluded that there is no correlation between project size and optimism bias, however there is a strong relationship between project size and the number of project risks. Major projects, like those included in the

Table 2.1 Optimism bias

Project type		Optimism bias %	
		Works duration	*CAPEX*
*Traditional**	Non-standard buildings	39	51
	Standard buildings	4	24
	Non-standard civil engineering	15	66
	Standard civil engineering	34	44
	Equipment/development	54	214
	Outsourcing	N/A	N/A
	All traditional	**17**	**47**
*PPP/PFI***	Standard buildings	−16	2
	Standard civil engineering	No info	No info
	Equipment/development	28	No info
	Outsourcing	N/A	N/A
	All PPP/PFI	**−1**	**1**

* The optimism bias is measured from strategic outline case or outline business case.
** The optimism bias is measured from the full business case. The capital expenditure optimism bias is measured as a percentage of the contract price
Source: Adapted from Review of Large Public Procurement in the UK (2002) Mott MacDonald.

Mott MacDonald study and minor projects, approximately £10 million in value have the same number of project risk areas whose project risks need to be managed, however, the number of project risks within project risk areas increases with the size of the project. Among the recommendations of the Mott MacDonald reports are:

- An open approach to sharing the successes and failures of major project procurements, through internal and external seminars, papers, etc.
- Post completion, 1 year after completion and 5 years after completion audits to compare project out-turns against projections, together with wide dissemination of lessons learned.
- Methodical archiving of key project documents.

Similarly, Flyvbjerg *et al.* conclude that in the public sector improved risk management could be achieved through:

- improved information transfer and communications,
- greater accountability.

Mega projects have their own particular type of risk and the analysis of risk in these projects is usually centred on Cost Benefit Analysis (CBA), a technique with a chequered reputation for accuracy and usefulness. Mega project risks may be catagorized (Flyvbjerg *et al.*, 2003) into the following groups:

- Estimates of earning capacity form the completed project, that is, numbers of projected passengers, cars, etc.
- Environmental risk – viewed by some experts as the most high profile risk at present.
- Regional and economic growth predictions, (i.e. the number of new jobs, etc.) that will be attracted to an area on completion of a project.
- Specific risks, such as multiple stakeholders, for example, government, global interests, special interest groups, etc.

Risk analysis and cost benefit analysis (CBA)

Although the idea of CBA is simple, the execution of a meaningful study can be problematic. CBA presumes that goods have a monetary value and is carried out by the assignment of a monetary value to each input (cost) and each output (benefit) resulting from a project. The value of the costs and benefits are then compared and in basic terms, if the benefits exceed the costs the project is deemed to be worthwhile. Simply determining what to include as a cost or a benefit can require careful consideration and may be subject to a difference of opinion. In addition while some inputs and outputs have stable and easily sourced costs, others may be more problematic or even change as a result of the project.

Many of these objections and difficulties are evident in real projects. One obvious example is the decision to proceed with Heathrow Terminal 5, announced in 2002. During the long public enquiry into the planned extension of the airport, many conflicting arguments were deployed about the costs and benefits to the local economy, environment and quality of life. A formal environmental impact assessment was undertaken, since this is required by law. But the continuing argument among supporters and opponents shows a lack of agreement about which costs and benefits are important and how they should be compared. In addition it should be recognized that decisions regarding projects will not and should not be made simply on the basis of CBA as this statement from the public enquiry demonstrates: *'unless Heathrow is able to maintain its competitive position there must be a substantial risk that London's success as a world city and financial centre would be threatened. By ensuring the continued success of Heathrow, Terminal 5 would make a major contribution to the national economy.'*

A typical approach to a CBA is as follows:

1. Clarify the issues, for example, what costs and benefits will be included and for whom the study is being carried out.
2. Identify the alternatives, including the do nothing option; this is often used as a benchmark.
3. Set out the assumptions. Assumptions are clearly part of the analysis and some will be better than others. It may be necessary to use assumptions for a wide variety of things, such as quantities, cost, duration of interest rates, etc.
4. List the impacts of each alternative project.
5. Assign values to the impacts. Appropriate monetary values should be attached to each of the impacts. For example, suppose that congestion, more intensive road repairs and increased accidents occur during the construction of a project. This will result in costs external to the project – for local authorities, hospitals, other drivers and pedestrians that will not normally be compensated. If one can value these, they can be added in, say, they amount to £10 million per year for the first 2 years. Similarly, external costs could occur during operation – for example, due to radioactivity from a nuclear facility or acid rain from a coal-fired power plant.
6. Discount future values to obtain present values, using techniques described in Chapter 3.
7. Identify and account for uncertainty and risk. Typically most aspects of a project are subject to uncertainty, for example, construction cost estimates. Some attempt must be made to

recognize risk and this may be as simple as offering a sensitivity analysis, that calculates the value of the project under differing outcomes or as complex as, a real options analysis that attempts to calculate the explicit value in light of the risk.

8. Compare the costs and benefits. The results are compared to determine if the project should proceed.
9. Post-project analysis. Important to determine the quality and the accuracy of the original analysis.

Mega project postscripts

In May 2003, David Black, an Edinburgh architect and author of a controversial book criticizing the Scottish Parliament project submitted a 20-page dossier to the European Commission with a request to investigate his allegations of mismanagement of the project, with the accusation that '*Scotland is becoming a laughing stock*' and blaming pre-devolution English civil servants for the 'ridiculous project'!

And finally the ultimate test of optimism bias, risk management and other theories relating to the disappointing performance of mega projects will be put to the test if the UK bid to host the 2012 Olympic Games is successful. Here is a case study of the story so far...

An example of how decisions are taken to go ahead with a mega project is illustrated by the following recent example. Although the risks are known to be great, to a large extent are unknown and unquantified and the financial investment considerable, the project proceeds, played out like a high stakes game of poker.

Cost benefit analysis and London's bid for the 2012 Olympic Games

First, comes the bluff; Lord (Sebastian) Coe, former Olympic medallist, speaking in the debate on the British Olympic bid in the House of Lords on 19 December 2002 said, '*It is time for the Prime Minister to be brave and for his Chancellor to see the big economic picture and to look at these figures properly. Only our ability to doubt our ability can hold us back.*'

There can be fewer bigger mega projects on the horizon in the next 10 years than the Olympic Games. On 15 May 2003, The Rt Hon. Tessa Jowell MP, Secretary of State for Culture, Media and Sport announced in the House of Commons the UK government's decision to back the bid for the 2012 Olympics with the statement: '*As the jointly commissioned ARUP report shows that we can deliver a high quality and competitive bid, based around an Olympic zone located in the Lee Valley.*'

The stakes were raised on 21 May 2003 when France announced that Paris would also be entering the race for Olympic gold and submitting a bid!

If the Secretary of State's statement is an accurate reflection of the information at her disposal, it appears that the decision to commit £13 million to bid for the Olympics, not to mention the £1674 million to build and stage the Games, should the UK bid be successful, was to a large degree, taken on the findings of a report commissioned by the UK Olympic Stakeholders Group comprising; the Government, the Mayor of London and the British Olympic Association and prepared by ARUP in association with Insignia Richard Ellis. The study published in May 2002 has been widely referred to by politicians and media sources as a 'CBA', see the earlier definition, despite the fact that the authors themselves admit in the body of the report that '*this appraisal is a hybrid between a cash flow business plan and a conventional CBA.*' Regardless of this health warning *The Daily Telegraph* reported on 27 May 2003 '*No British bid has been better prepared … much time and effort has been put in making sure the figures are right.*'

The ARUP cost benefit analysis was prepared on an outline proposal for a 'specimen' Olympic Games for appraisal purposes and without consultation with many of the potential main players. The specimen games including such facilities as; an Olympic Village, main stadium, warm-up track, etc. By the admission of the authors of the report, there was an absence of detailed risk assessment and instead a sensitivity analysis was used to estimate overall risk.

Sensitivity analysis

There is usually a wide range of proposed numbers attached to any project. Estimates of the amounts of materials and labour required, prices for these inputs, the number of final users, what value to attach to their use and an appropriate rate to use in the discounting of future costs and benefits, may all be subject either to error or disagreement. The most responsible response to having ranges of estimated or suggested values is to offer calculations based on various scenarios and to discuss the effect that different numbers have in a sensitivity analysis. That is, one should clearly state the effect of changes in various values on the final assessment of the project. For example, if construction costs are 10% over the estimate, how will the net benefit of the project change? If a higher interest rate is used to discount future costs and benefits, will the project still be feasible? One common approach is to calculate best case, worst case and intermediate scenarios. That is an analyst would first calculate the net present value of a project using the values that will maximize its value, then the values that will minimize its value and finally some middle values. This will give the policymakers an idea of the degree of uncertainty surrounding the project and how important it might be.

In the Olympic bid the elements thought to be at most risk were identified and the study concluded that there was risk of both substantial cost over-run in capital works, (i.e. building the various stadia and sports venues, as well as reduction in income). The summary of the financial analysis is shown in Table 2.2.

Table 2.2 Cost benefit analysis

	Expenditure (£m)	Income (£m)	Surplus/ deficit (£m)
Bidding for the Games	13	7	−6
Operating the Games	779	864	+85
Sports development programme	167	0	−167
Capital cost of infrastructure and facilities	403	0	−403
Land purchase and residual value	325	431	+106
Cash flow balances	**1687**	**1302**	**−385**
Provision for risk	109	–	−109
Cash flow including risk	**1796**	**1302**	**−494**

Notes: All figures based at 2002 prices discounted at 6%.
Source: ARUP, London Olympics 2012: Costs and Benefits.

The cash flow calculations include £109 million provision for risk. The basis for the figure was the assumption that the bidding costs would increase by 5% and the capital costs of constructing the Games infrastructure will be between 30 and 50% higher than the £403 million included in the study! In addition to these risks the report also identifies that other risks associated with transport proposals and security. No figure or probability was given to the chances of these risks occurring, but the report concluded that these risks will need to be taken into account, but did not suggest when, or by whom and should be set against the opportunities to avoid or mitigate risk through management, anticipation and planning. The final conclusions of the report, with other factors taken into account, are shown in Table 2.3.

By way of emphasizing the points made earlier about the difficulties of using CBA, it can be seen from Table 2.3 that the result of the exercise can be almost anything that you want it to be. A criticism levelled at the public sector comparator when used to evaluate PFI projects is discussed in Chapter 4. In this case bridging the £494 million shortfall in costs (expenditure) and benefits (income) with a windfall of between £280 and £610 million from additional tourism. Given the impact on tourism of the September 11th incident and the widely predicted future acts of global terrorism between now and 2012, the questions must be asked, how is this additional income possible? And what is the basis of this optimism?

To many within the construction industry the spuriousness of these figures is breath-taking, perhaps when it comes to

Table 2.3 UK Olympic bid

	Expenditure (£m)	Income (£m)	Surplus/ deficit (£m)
Cash flow including risk	1796	1302	−494
Additional tourism income	103*	280–610	+280 – +507
Other quantified benefits	0	69	+69
Total cash flow including benefits	**1899**	**1651–1981**	**−145 – +82**

*Additional investment to secure additional tourism for higher income forecast.
Source: ARUP, London Olympics 2012: Costs and Benefits.

risk, mega projects are a separate breed of construction project whose performance should be judged by different criteria?

Non-mega projects

In an attempt to reduce exposure to risk, consolidation by acquisition and merger is increasing across design, construction and supply chains. Many household names both on the supply and the demand side of the construction process are increasingly seeking the reassurance that comes from a large multi-disciplined organization. The next few years are expected to see this trend continue resulting in the emergence of the so called 'A Teams' in design, engineering, construction and supply. The advantages for the larger organizations are that they can:

- carry risks more effectively because of the range of skills and expertise within their organizations,
- insure against those risks they no longer wish to carry.

Unfortunately, this trend could leave the rest of the industry on both the demand and the supply side left with the 'B' and 'C' teams.

A glance at stock market performance, world-wide over the last 10 years and in particular since the burst of the dot com bubble in 2000 demonstrates how little room for manoeuvre there now is for many organizations. Large corporate clients and the insurance companies that have traditionally insured these organizations simply do not have the balance sheets to consider high risk projects or high risk procurement strategies in the current climate. Instead clients and their insurers are demanding procurement with less risk, more certainty and the opportunity to take out extended warranties, etc., an area that will be discussed later in this chapter. See also Chapter 7 – procurement in France.

Consequently, the insurance sector has had to realign its approach to risk. Traditionally, global insurance companies have lost money on insurance underwriting, however as long as the stock market was buoyant, companies were able to off-set insurance losses with stock market profits. However, with the dramatic downturn in the performance of stock and shares

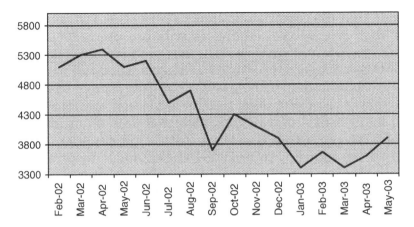

Figure 2.1 FTSE 100 share index.

world-wide in the early part of 2000, see Figure 2.1, insurance companies found themselves very exposed to losses from their insurance business and as a result began to rethink their business strategies in this sector. Towards the end of 2002 construction firms across the UK experienced increases in insurance premiums ranging from 200% to as much as 500% across a whole range of insurance classes from employer's liability to professional indemnity. In addition competition in the insurance market was drastically reduced in the late 1990s when three large insurance companies merged and a few years later Independent Insurance, one of the most popular and cheapest insurance companies in the marketplace stopped trading. One event above all others however was to have a massive impact on the way in which risk is viewed – namely the terrorist attacks in New York on 11 September 2001. The destruction of the World Trade Centre made an increasingly nervous insurance sector – very nervous indeed. This hit reinsurance firms – the companies that insure the insurers – with big losses. The reinsurers then passed these costs to the underwriters who in turn transmitted them to construction companies. By 11 March 2003 the FTSE 100 index had halved its value since the close of business on 31 December 2002. On the same day the Japanese stock exchange closed at a 25-year low.

In addition to the above factors other trends had been contributing to an insurance crisis. British society is growing increasingly litigious – some industry commentators are of the opinion that the construction industry gets the insurance products that it deserves, adding that if procurement practices change, then it will get new insurance products, however in the meantime adversarial methods will get adversarial insurance. The problem is that if the system requires a proof of liability, that splits the project team into its individual members – each with its own firm of solicitors, about 65–75% of an insurance payout goes into claims related costs. This situation has led some of the industry to promote single project insurance and this will be discussed in detail later in this chapter.

The consequences of ignoring risk can be truly humiliating, not to say career threatening. History is littered with contractors; some of them household names like Laing, for example, who did not see the elephant coming, see Figure 2.2, and signed contracts that contained too much risk, or risk that the contractor was unable to manage. Laing Construction, the construction arm of John Laing was sold for £1 in October 2002 to O'Rourke, one of its own sub-contractors. Laing Construction had been established for over 150 years and was responsible for the UK's first motorway,

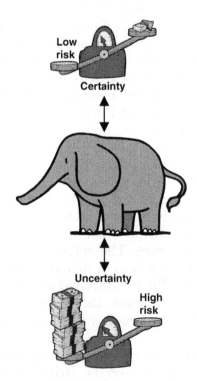

Figure 2.2 Recognizing risk.

the M1 and Coventry Cathedral. However, two disastrous con-tracts, the £116 million PFI National Physical Laboratory proj-ect, where Laing under-estimated the risk contained in the performance specification and was unable to achieve the required temperature control of rooms, which were required to vary by less than half a degree, and in which Laing lost £40 million. This proj-ect that is currently being investigated by the National Audit Office, and further big losses (£32 million) on the suicidal bid for the Millennium Stadium in Cardiff. Problems with the Cardiff project involved the redesign of the retractable roof, increased material costs, changes to demolition plans and poor weather and led Laing to first announce in October 1998 that it was pulling out of contracting, that is to say competitive tendering and finally put itself up for public auction in 2001.

Nowhere are the differences between the traditional procure-ment systems and the so-called relationship contracting and PPP procurement discussed in Chapters 4 and 5 so apparent than in the area of risk. The traditional approach to identifying the relationship between risk and procurement strategy (Flanagan and Norman, 1993) is illustrated in Figure 2.3. Within this framework the relationship between risk and procurement systems lies within the area of risk response, that is, which pro-curement path will best suit the perceived risks.

This risk management framework illustrates that in trad-itional procurement the risk identification/analysis is used by the client and professional advisors to choose the correct approach. Only after the perceived risks have been identified

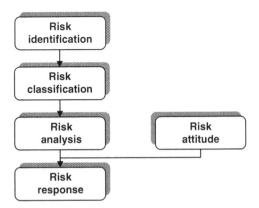

Figure 2.3 Risk and procurement strategy (source: Flanagan and Norman, 1993).

and without reference to the supply chain is the decision made. Whereas in relationship contracting, see Chapter 5, the input of the supply chain/prime contractor or alliancing partners are vital parts of the procurement process, involving the mutual identification of risks and deciding which party is better able to manage and control them.

Risk is like an elephant – hard to describe but you recognize it when you see it! The Royal Institution of Chartered Surveyors warns in its publication *The Surveyors' Construction Handbook* (1998), that *'the production of a new construction project is seldom straightforward, being subject to numerous risks and uncertainties.'* The question that construction clients surely must ask is – why? Why is construction different from other market sectors and why are construction clients asked to be exposed to so much risk? The traditional picture of risk in the procurement process is of two sides; the client and the contractor sparring with each other, trying to off-load as much risk onto the other party without them realizing, until it is too late, and the contract has been signed. The 1980s was characterized by clients attempting to transfer more and more risk to contractors, through such devices as management style procurement paths and/or bespoke contracts conditions. History shows many of these innovative contracts became unravelled in the 1990s to the delight of lawyers.

At the start of a construction project the client owns the risks that are inherent to that project. The success of the project depends in part upon how effectively these risks are managed by clients and their advisors. In broad terms the client can choose from a risk transfer approach at one end of the spectrum to a shared risk approach at the other end, with a variety of risk sharing hybrids in between. It cannot be said that any particular model along this spectrum is the right one and even the most high profile clients, despite the PR, run a variety of procurement strategies. What is clear however is that there is a strong swing away from traditional approaches to procurement to approaches that embody partnering and collaboration. Clients and their advisors must choose the most appropriate to the particular circumstances of the project.

Why have clients traditionally sought to transfer risk to contractors? The simple answer is that they have had to in order to ensure that their own position is more predictable in terms of project cost, completion dates and return on investment. Clients are demanding more and more innovative solutions from contractors, because of their dissatisfaction with traditional procurement systems and uncertainty surrounding the attainment of objectives. Of course in between total certainty and total uncertainty, if there are such things, there are numerous levels of risk capable of having a substantial impact on project performance.

Bearing in mind the fate of Laing why do contractors enter into contracts with excess risk? The answers are quite simple:

- They have to in order to maintain cash flow, traditionally profit margins for UK contractors are low, 3–4%, coupled with the belief that adversarial contracts and industry culture will enable them to ride out the storm.
- UK contractors are in the main a one-product organization they cannot afford to pick and choose. In comparison Bouygues, the French giant, established in the 1950s has a wide diversity of business interests including telecom and the media.

Also it is clear that in using traditional procurement systems the successful bidder will be the contractor who is:

- most optimistic about cost uncertainties,
- most optimistic about the certainty of claims occurring during the contract period,
- most desperate for work and therefore is prepared to pare profit margins to the bone,
- least concerned about reputation and bankruptcy,

an approach that carries a high degree of risk.

Risks will vary during the contract, the procurement phase being one of the most prone to risk. How can the risk be managed? Risks can be classified into internal and external:

- Internal risks are those found within the project, and are more likely to be controllable but on the other hand can vary between projects, for example, construction delays attributable to a number of factors.
- External risks are those generated outside the project and in many circumstances are uncontrollable – they are external to the market. Such as political risk – during the duration of a PPP project, for example, a government may

change two or three times. In construction projects externalities will be positive as well as negative. Perhaps one of the most important risks in terms of impact at present and for the foreseeable future will be environmental risks.

The processes involved with risk management are:

- Risk analysis,
- Risk management.

Risk analysis

Risk analysis takes place over a range of different items identified as areas where uncertainty is likely to impact on the planned progress of the project. For example, in the case of a contract requiring a large amount of bulk excavation, what is the likely impact of inclement weather on the work and on the remainder of the project and who is to assume responsibility for this risk and how will it be allowed for? Traditional forms of contract, such as the JCT (98) have, when used, allocated the risk between the various parties to the contract. It was then traditionally the role of the contractor to shed as much of this risk as possible through the supply chain, that is to say subcontractors and suppliers. Traditionally, quantity surveyors, even the great and the good, tried to mitigate the impact of risk by including contingency sums within the bills of quantities, both in the form of provisional sums, as well other hidden 'pockets' of money scattered throughout the measured work sections, the existence of which were known only to the professional who had produced the bills of quantities and could be produced, like a rabbit out of a hat to compensate for the impact of risks. In the traditional approach to procurement each party has a different list of objectives.

Risk management

Risk management generally involves the use of quantitative and qualitative evaluation techniques. Most projects do not enter into qualitative analysis but virtually all projects make use of qualitative risk analysis. Risks do not impact upon a project individually but in combination, yet there are few if any tools available accurately to model this.

Quantitative techniques: modelling risk in order to quantify its combined effect on the project using,

- decision trees;
- influence diagrams;
- Monte Carlo simulation;
- sensitivity analysis, described earlier.

With these techniques the likelihood of risk occurring is given a numerical probability, for example,

$$0 = \text{impossible for risk to occur,}$$
$$0.5 = \text{even chance of risk occurring,}$$
$$1 = \text{risk will occur.}$$

Qualitative techniques: identifying, describing, assessing and understanding risk. This can take the form of

- a risk register, see Table 2.4;
- a descriptive statement of relevant information about a potential risk.

Even in collaborative/partnership relationships, risk cannot be ignored. For example, a risk register forms an important part in PPP/PFI bidding process.

Risk register

Many standard forms of contract now make risk analysis an obligatory part of any procurement process. One of the more common ways to identify and allocate risk is with the use of a risk register or matrix. One of the advantages of this approach is that there can be no doubt as to which party has responsibility for what risk. Table 2.4 illustrates a typical risk matrix for a PPP project; a similar approach is just as appropriate to the private sector. For many sectors using PPP/PFI procurement the allocation of risk is by now fairly well known.

As discussed later in Chapter 4 appropriate risk allocation between public and private sectors is thought to be the key to achieving value for money on PPP/PFI projects.

Table 2.4 Risk register

Description of risk	Included in PSC?	Public sector risk	Shared risk	Private sector risk
Demand and revenue risks				
Type of service demanded	√	√		
Asset usage	√		√	
Capability of assets	√			√
Operating risks				
Operating costs	√			√
New legislation	√		√	
Health and safety costs	√			√
Exchange rates	√			√
Integration of new equipment	√			√
Design and construction risk				
Site availability	√		√	
Commissioning	√			√
Site costs	√			√
Design errors	√			√
Late design changes	√		√	
Weather	√			√
Strikes	√			√
Noise restriction	√			√
Construction defects	√			√
Unproven technology	√			√
Contractual claims	√			√
Fire	√			√
Health and safety	√			√
Other risks				
Availability and cost of finance	√			√
Taxation	√			√
Legislation	√	√		
Inflation	√		√	

Examples of project risk areas that require management during the procurement process are due to:

- *Complexity of contract structures*: This may include, unclear details of risk allocation and/or payment mechanisms, as well as protracted negotiation on contract terms.

- *Late contractor / supply chain involvement in design*: Involvement by the contractor supply chain as early in the design/procurement process will help ensure 'baked in' buildability. Late involvement precludes involvement in value management.
- *Contractors' capabilities*:
  - Do not select contractor on the basis of availability and price alone,
  - Rigorous pre-qualification procedures for contractors and their supply chains, see, National Health Service ProCure 21 system discussed in Chapter 5, Relationship Contracting, will ensure that only contractors and supply chains with proven competencies over a range of critical criteria are considered.
- *Disputes and claims*: Ensure that there are established mechanisms effectively to manage adversarial relationships between project stakeholders.
- *Information management system*: Project stakeholders need to be able to communicate in real time.

Given that risk can never be totally eliminated, the next considerations are the ways by which a client/procurement team can insure that the effects of risk are mollified.

Project insurance/inherent defects insurance

As mentioned previously in this chapter, not only are clients demanding more certainty in construction procurement but also they are seeking to mitigate risk by demanding long warranties for the completed built asset. The lack of project or inherent or latent defect insurance cover in the UK construction industry has been cited from many sides, (Goodall, 2001; Robinson, 2001) as a major barrier to delivering value for money. The argument is that the absence of latent defect insurance, in essence a guarantee, perpetuates the ethos of silo working with every man/woman for themselves, in the case of defects which appear during the first years of occupation and use of a new building. An inherent defect is a defect that exists, but remains undiscovered, prior to the date of practical completion but later manifests itself by virtue of actual physical damage which may not have been reasonably discovered previously. Under the present insurance system of separate Professional Indemnity and Contractors All Risks, the consultants and contractors respectively, the issue of defective workmanship is not covered. However, the contractor is the party essentially responsible for workmanship and so liability rests with the builder. The tension rises out of the fact that the contractor cannot contribute to the design and could feel that buildability and quality could be a problem within the project but there are no means whereby they can effectively be involved in such issues. The current system can therefore be seen to not deliver value or security to the client. The origin of the defect could be; poor workmanship, unsuitability of materials or poor design, structural failure, etc. A number of insurance companies in the UK are now writing policies to cover the cost of remedial work to correct latent defects in finished construction projects for a period of 10–12 years after practical completion. The policy is written in the first party and therefore in order for the policy holder to benefit from the scheme negligence does not need to be proved prior to remedial work being carried out. Unlike France, where 10-year defects insurance is a compulsory legal requirement, see Chapter 8, it is estimated that less than 10% of all UK projects carry latent defects cover. Instead in the UK a system which operates at a number of levels; professional indemnity insurance, collateral warranties, employers liability insurance is preferred by many.

According to professionals active in the field, the difficulties facing insurers and preventing the wider use of latent defects insurance are as follows:

- lack of statistics,
- lack of experienced underwriters,
- the potential cost of failing to underwrite properly,
- difficulties in defining the cover,
- lack of demand,
- the poor image of the UK construction industry with the lowest productivity and skills levels of all European States. Low labour costs tend to equate with low productivity.

The proponents of latent defects insurance argue that the presence of latent defect insurance helps to promote an environment where the design team operates in an non-adversarial relationship and in addition helps to promote buildability and predictability. In addition the present system of professional liability reserves the function of the design to the employer's professional advisors (architects, engineers, etc.) and reduces the function of the contract to that of the simple execution of drawings, specifications and instructions.

Latent defects insurance has been available in the UK for about 20 years but only became popular from 1989 onwards when a number of UK owned insurers brought out their own policies. Until recently the cover was limited to the repair of damage to a building caused by an inherent defect in the main structure of the building. More recently the extent of cover offered has been widened although many believe that cover remains quite limited. It is common practice in the UK for policies to have excess, that is a proportion of any claim to be paid by the insured, of £25,000 for structural elements. The system is monitored by a series of technical audits. These audits are an essential feature of the Inherent Defects Insurance and are common to other latent defects of guarantee schemes that operate elsewhere in the world. An independent consulting engineer or mechanical and electrical engineers carry them out. Their roles are to monitor the project throughout the design and construction stages on behalf of the insurance company in order to minimize their risk. They report as the work progresses that the project is being constructed to a normal standard in accordance with good building practice. The technical auditors:

- Monitor the design and construction method.
- Check that there is a clear demarcation of responsibilities within the design team.
- Visit the site at regular intervals.
- Issue a certificate of approval to the insurance company at practical completion.
- Carry out triennial audit health checks on mechanical and electrical installations.

In practice the range of situations encountered by the technical auditors range from the discovery of errors in the calculations for components to works on site by the contractor that contravened good practice.

The basic inherent defects policy covers actual physical damage to the whole of the building, but importantly, only that caused by inherent defects that originated in the structural elements, such as foundations, external walls, roof, etc. In addition, optional cover can be added for non-structural parts such as, weatherproofing and mechanical and electrical services. If the mechanical and electrical option is selected then in some cases the failure for the completed installation to meet the performance specification will be covered. Finally, fitness for purpose cover can be included, for example, minimum internal floor areas.

The policy covers the following items:

- the cost of repairing damage in the main structure,
- the cost of remedial action to prevent imminent damage,
- professional fees,
- cost of debris removal and site clearance,
- extra cost of reinstatement to comply with public authority requirements.

The policy is a first party material damage policy which essentially means that there is no need to prove negligence by a third party and cover may be freely assigned to new owners, lessees or financiers.

The benefits for construction projects and clients if latent defects cover is in place are:

- There is no need to rely on the project team's Professional Indemnity Insurance.
- The potential for confrontation is reduced.
- There is more peace of mind for all the project team.
- Less time and money is spent on arguing about contract conditions and warranties.
- Innovation is encouraged.
- Everyone can concentrate on getting the actual design and construction right.
- Can be a major advantage when negotiating a sale or letting.

So, what does this peace of mind cost? According to insurance companies AON and Royal SunAlliance the basic structural cover plus weatherproofing for 12 years can be obtained at rates ranging from 0.65 to 1% of the total contract value, rising to £2% for completed buildings. Total cover including all the options can cost from between 1% and 2%. The premiums are usually paid by instalments and include the fees of the technical auditors described below. For example, for a project with a contract price of £2 million cover could cost up to £20,000.

The current system is a complex arrangement of insurance cover effected by professional indemnity insurance, and large numbers of collateral warranties, typically 20–30 for a construction management project, for example, that reinforces the silo working mentality as well as discouraging contractors to engineer waste out of the working details and to introduce buildability. It creates a safe commercial environment where team working and, sharing of ideas can help reduce conflict and risk.

Subrogation

An interesting and very British variation on the latent defects cover is subrogation. As previously mentioned this means that liability does not have to be proved in the case of a claim for defective workmanship etc. However, there is a significant impact upon the project team in relation to this issue. A definition of subrogation is 'the right of an insurer (the insurance company) to stand in the place of an insured (the client) in order to exercise the rights and remedies which the insured has against third parties for the partial or full recovery of the amount of a claim.' In other words the insurance companies may pursue what they perceive as negligent members of the project team in order to mitigate their losses. The client will avoid this route, but not so the project team and therefore may be subject to a second wave of litigation. What is the incentive therefore for the team to integrate and more specifically for the contractor to innovate and contribute to the design?

A way of averting the above situation is through the use of subrogation waivers. In this event the insurance company, for the payment of additional premium, will not enforce its claim against say an architect, surveyor or engineer and/or contractor in the event that the latent defect is caused by their negligence. To some it is thought that if waivers are not used then there is little or no likelihood of eradicating the back-watching, trail covering adversarial culture within which the design consultants and the contractor will often find themselves operating.

A move towards less prescriptive procurement documentation

The traditional approach to construction procurement where a lump sum contract is selected (still used in 45% of all construction procurement), will normally be centred on a bill of quantities, drawings and other contract documents. The bills of quantities, which become a contract document, are deliberately very prescriptive, usually being divided into sections typically:

• Preliminaries,
• Preambles,
• Measured items.

The first two sections set out in unequivocal terms how the successful bidder will construct the new project, laying down process, procedures and standards. The prescriptive nature of the traditional bills of quantities is to produce a document with which identical bids can be submitted by contractors or sub-contractors and in addition to allow the easy analysis and evaluation of bids by quantity surveyors. For example, the following extract is from the Masonry Preambles section of a bill of quantities produced for a local authority, circa 1986:

Mortar
1. Measure materials using clean gauge boxes.
2. Mix ingredients thoroughly to a consistency suitable for the work and free from lumps.
3. Keep plant and banker boards clean at all times.

and so on for 176 pages, for a contract with a value of less than £0.5 million!

Even in countries where quantity surveying is not practised as a separate profession, a bill of quantities or schedule of rates, usually forms part of the bidding process, although not usually a contract document and not prescriptive.

Following a review in 2002 of the National Building Specification (NBS), an organization used by many professionals as the basis of project specifications, has been removing specifications of construction processes from its standard clauses, that is to say making the contract documentation less prescriptive.

Process constraints still have validity, for example, restrictions on working hours, noise, etc. and would normally be included in the preliminary section of the bills of quantities. However, it is becoming apparent that there are few good reasons for wanting to describe the construction process themselves. Except perhaps, where unique or highly specialized products or services have to be sourced. On the other hand, there are several good reasons for not wanting to describe processes including:

- Specifying on autopilot using habit and tradition.
- Concerns about buildability and health and safety.
- Overconfidence of the specifier, convinced they know better than the contractor.
- Lack of confidence in the contractor's ability to carry out the work.
- Losing sight of the outcome of deliverables.

Reasons for not specifying processes

The designer

According to NBS, choice of specification method is rarely a conscious, rational decision. Usually it is down to habit, with no real rhyme or reason. Accordingly, logical use of process specifications is unlikely as:

- Specification of process undermines the claim that architects and their kin do not supervise. Deliverables can be checked at any time after execution, the process cannot.
- Specifying processes blurs the distinction between the roles and the responsibilities of the designer and the contractor. Some roles and responsibilities may be unallocated and overlooked.
- Describing the outcome and the construction process means the describer is taking the responsibility for the design and construction of the item, in addition locking the contractor into a particular process means that the specifier assumes responsibility for health and safety issues.
- Often processes are described without the required result, in a vacuum.

Lack of trust

- Competent builders will resent being told how to proceed – lack of value demonstrated in builder's professionalism. An important part of competition is the choice of construction processes; it encourages innovation and competitive advantage to the detriment of the client. Changing the specification through substitution is difficult for the contractor and is often resisted in principle by the professional advisors, particularly with the system of liability operating in the UK. However, as discussed in Chapter 8 other countries encourage contractors to submit alternative solutions, often incorporating say, time efficient processes for cost efficient processes.
- It is often easy for the contractor to go with the specified process. That is, the contractor will not evaluate the specified process, will not consider alternative processes and may even stop thinking about the process altogether.

Outcome

- Even strict adherence to a sound process will not guarantee that the desired outcome will be achieved as the chance of a successful outcome diminishes when the specified process is not sound incomplete, out of date, erroneous, etc. for example, the use of High Alumina Cement in concrete in the 1970s.

What is the alternative to prescriptive specification?

As well as describing outcome or deliverables other approaches to be considered are:

- *Specify the deliverables, outcomes*: Performance specifications have been in use, particularly in the mechanical and engineering sectors, for many years. However, the spread of PPP/PFI projects has led to non-prescriptive specifications, that are written in output terms being more widespread in other sectors of the construction industry too.
- *Pre-qualification of tenderers*: Ensure that the builders are inherently competent, rather than being chosen on the

basis of size and availability. Process specification will not compensate for incompetence, see Chapters 5 and 6.

* *Submittal of method statements*: Required at bid stage for critical parts of the work or during the contract.

Conclusion

The mismanagement of risk has, and continues to be, the primary cause of cost and time over-runs in the construction industry. Risk will never be eliminated from the construction process, nevertheless there would appear to be a class of mega projects that will continue to ignore the impact of risk on project outcomes. However, as demonstrated by this chapter, for more modest construction projects there are approaches that can successfully ensure that the procurement process is not simply a matter of passing risk through the supply chain like a hot potato, until someone has the nerve to get their fingers burned.

Bibliography

Arrow K.J. and Lind R.C. (1996). *Cost–Benefit Analysis: Risk and Uncertainty*, 2nd Edition. Cambridge University Press.

ARUP (May 2002). *London Olympics 2012: Costs and Benefits*. ARUP, London.

Bose M. (2003). Capital way to win votes. *Daily Telegraph*, 27 May, p. S4.

Flanagan R. and Norman G. (1993). *Risk Management and Construction*. Blackwell, London.

Flyvbjerg B. *et al.* (2003). Mega Projects and Risk: An Anatomy of Ambition. Cambridge University Press.

Mott MacDonald (July 2002). *Review of Large Public Procurement in the UK*. HM Treasury.

Richards M. (2002). The fight for survival. *Building Magazine*, 22 November, issue 45.

Richards M. (2002). Insurance crisis to bring sites to a halt by Easter. *Building Magazine*, 22 November, issue 45.

Web sites

www.olympics.org.uk/index.asp

3

Procurement –
a building is for life

Introduction

This chapter is concerned with the consideration of procurement within a context of sustainability, whole life cost, facilities management and current UK taxation regimes. How many companies combine finance with management, procurement, operations and facilities management?

Sustainability – fashion or here to stay?

Various attempts have been made to define the term 'sustainable construction'. In reality it would appear to mean different things to different people in different parts of the world depending on local circumstances. Consequently, there may never be a consensus view on its exact meaning; however, one way of looking at sustainability is *'The ways in which built assets are procured and erected, used and operated, maintained and repaired, modernized and rehabilitated and reused or demolished and recycled constitutes the complete life cycle of sustainable construction activities.'* Given this definition therefore it is vital that sustainable issues are given due consideration as early as possible during the procurement process.

Since the Industrial Revolution, the world has witnessed incalculable technological achievements, population growth and corresponding increases in use of resources. As we enter a new century it is recognized that the success of technological advances are; pollution, landfills at capacity, toxic waste, global warming, resource and ozone depletion and deforestation. These factors are straining the limits of the Earth's ability to provide the resources required to sustain life while retaining the capacity to regenerate and remain viable. Research establishments aided by private practice, are in the course of assembling data on the impact of unregulated industrial growth which is generally accepted to be huge. For example, the construction sector is responsible for one-sixth of the total freshwater withdrawals and taking into account demolition, generates 30% of waste in OECD countries. In addition around 40% of total energy consumption and greenhouse gas emissions are directly attributable to constructing and operating buildings according to the Energy Efficiency Best Practice Programme, recently rebadged; Energy Action. Measured by weight, construction and demolition activities also produce Europe's largest waste stream, between 40% and 50%, most of which is recyclable. Clearly then given the scale of the consumption of resources by construction, even small percentage improvements in building efficiency would translate into many billions in savings. In absolute terms the challenge to become more sustainable in the long term is a greater challenge than that of any other sector. As will be discussed in this chapter, the most effective time to introduce/consider sustainability issues are at the procurement stage.

Sustainability is more than avoiding another energy crisis, as typified by the sustainable environment crusade of the 1970s, when road speed limits were reduced and solar power became popular. However, as quickly as the capital and cultural boom of the 1980s gained momentum concerns over energy efficient buildings showed a marked decline, as other priorities took centre stage. Despite this change of emphasis, the underlining facts remain; the choice of building systems, materials and design during the procurement of built assets impact on the environment through; global warming, the production of atmospheric carbon dioxide and carbon emissions. In addition, the energy consumption associated with material production and transportation impacts on the depletion of gas, oil and coal reserves. However, the energy consumed in extraction and transportation of materials, etc., is only a fraction of the energy used over the life of the building. Data on so-called 'embodied energy' is limited as it is only fairly recently that this aspect has been considered; however the main consideration here is to maximize the efficiency of the design so as to reduce the energy demands over the life of occupation; for example, refrigerants in occupation. The burning of fuels, coupled with the chemical processes to refine construction materials, emits pollutants and toxins into the air, as well as adversely affecting the ozone layer. As will be discussed later techniques such as whole life costs and life cycle analysis can now help select the most appropriate materials, etc.

What has sustainability to offer construction?

There follows a number of case studies of the application of sustainable processes to school projects kindly supplied by Atkins, Faithful and Gould.

Clients, especially public clients, must take the lead in promoting sustainability in construction procurement. Sustainability impacts and their mitigation should be addressed as far as possible in the planning and design stage, prior to commencing tendering. Attention to some general design principles very early in project development can influence sustainability profoundly. Attention to the following issues will increase the design costs but not the costs of the building itself, and will reduce whole life costs:

- Short supply chains to reduce transport costs.
- Exercise waste minimization and recycling in construction.
- Building orientation.
- Durability and quality of building components, generally chosen to last for the appropriate refurbishment or demolition cycle.
- Local sourcing of materials, particularly the heavy or dense ones.
- Design sensitive to local topological, climatic and community demands.
- Construction type – prefabrication, wood or concrete structure.

Attention to such fundamentals of design requires close collaboration between engineers and architects from the beginning of the design process and through the procurement phase. The renewable option should also be considered early in the procurement process too, but depending on the level of integration their addition may only have to be considered alongside the design of the services. Renewables, with some notable exceptions depending on local circumstances and government grants, will generally require a significant cost increase so the policy regarding renewables and the possibility of ring fencing a certain part of the budget for them should be investigated as early as possible. Another reason for considering the above issues early on, is to give time to apply and process grants. The DTI already gives grants for photovoltaics (70% for a school) and are now initiating a programme to award grants to biomass systems and solar thermal. If the site is particularly windy, then wind power could be an attractive proposition.

The case studies that follow are of new schools. Most options for a sustainable school, though, fall between the cheap and the expensive. Some measures considered a luxury a few years ago will be mandatory under the requirements of new Part L of the Building Regulations. Building costs will have increased and require better design, but this will result in lower whole life costs. Passive design to increase solar gain in winter and to exploit natural ventilation and daylight can sometime have cost implications, but have the benefit of simplicity and reliability over mechanical and electrical systems which save energy.

Although minimization of energy use plays a major part in these case studies because of its effect on global warming, sustainability also means attention during the design to selection of the most appropriate materials.

During procurement supply chains should be aware that components should be chosen selectively to minimize:

- Embodied energy, that is energy of production and transport.
- Atmospheric emissions, for example through the use of low NOx boilers and avoiding insulation whose manufacture results in phenol emissions.
- Disposal to landfill of non-biodegradable waste, for example by using organic materials or components which are recyclable.
- Air quality contaminants, for example solvents and wood preservatives continue to emit volatile chemicals long after construction, though in much smaller quantities, and these have been implicated in 'sick building syndrome'.
- Replacement due to poor durability.
- Use of finite resources, or at least promote the use of materials like wood from forests which are being replenished.

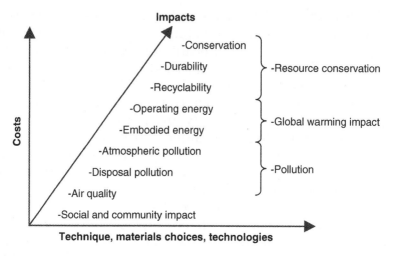

Figure 3.1 Techniques, materials choices, technologies (source: Atkins, Faithful and Gould).

When opting for sustainable choices the designer must commit much more time and therefore cost to design. Sourcing new local and recycled materials, and checking specifications for durability, is an overhead of sustainable design which pays off handsomely in buildings with lower whole life costs and reduced environmental impact. The process of getting the minimum whole life cost and environmental impact is so complex, being a three-dimensional problem as indicated by Figure 3.1. Each design option will have associated impacts and costs, and trade-offs have to be made between apparently unrelated entities, for example what if the budget demands a choice between recycled bricks or passive ventilation.

The solution to a complex problem like this will be iterative. Consultants will provide sustainability advice with a different focus and resolution at the inception, design and construction phases, to eliminate unrealistic options as soon as possible.

Case studies

Throughout this book heavy emphasis is put on moving away from the traditional confrontational relationships, so typical of construction procurement in the past, to inclusive, non-confrontational strategies based on trust. In the examples included below the aim to produce buildings that include

Figure 3.2 Lammas Secondary School, Leyton, East London.

sustainable features has only been possible by the adoption of integrated and collaborative procurement approaches.

1. Lammas Secondary School

The new Lammas Secondary School in Leyton, East London, is a private finance initiative (PFI) project based on the partnership between WS Atkins Consultancy, Wates Construction and Innisfree. It is the first secondary school to be constructed in the London Borough of Waltham Forest for 100 years and was constructed on a brown field site on the edge of Hackney Marshes on land formerly occupied by factory premises which have now been demolished (Figure 3.2).

W.S. Atkins design has been developed in close collaboration with its construction partners.

Sustainability design features include:

- extensive fenestration to optimize natural lighting,
- high thermal insulation,
- solar shading devices and passive ventilation systems to cool the building in hot weather,
- door entry control for all external access doors,

- L1 standard fire detection system (this is the highest category of protection),
- PIR detection to all ground floor and vulnerable first floor rooms,
- window shutters on IT rooms.

2. *Meadowside Primary School, Gloucestershire County Council*

The school was designed and built to sustainable principles and specification, with elements designed as sustainable education tools. The aim was to embed sustainability into the school ethos and curriculum (Figure 3.3).

Sustainability design features include:

- Encourage all-weather access on foot by providing oak porch for waiting areas, while car access was discouraged by design. Cycling encouraged by providing safe routes.
- Rainwater captured by porous paving and underground storage.
- Roof mounted wind catchers used for natural ventilation without heat loss.
- Sun pipes used to light interior corridors.

Figure 3.3 Meadowside Primary School, Gloucestershire County Council.

- Bamboo used for hall floor.
- Tierrafino Clay Plaster finish at higher levels.
- Visible metering used as a teaching aid.

3. *Dalkeith Schools community campus*

When a local authority in Scotland entered the PFI arena look-ing for a school, sports and community centre with leading environmental credentials, Atkins responded with a design that incorporated the following sustainability features:

- Low energy design, with a SEAM rating of A.
- A method for cooling the structure using a design solution known as TermoDeck®.
- Sustainable urban drainage techniques to help stabilize temperatures.

The challenge was to design an environmentally sound sec-ondary school for 2100 pupils, 100 special needs pupils, and third party leisure use. The project had to be delivered within tight financial constraints. The design brief called for the school to achieve a Schools Environmental Assessment Method (SEAM) rating of A.

To meet the SEAM rating and manage the space to the tightly defined temperature bands set in the output specification, the team investigated the use of TermoDeck® which pumps air through the floor slab. The concrete density provides a 'steady' temperature to keep the building cool.

The TermoDeck® system, when used with other design fea-tures such as external shading devices, uses the thermal mass of the building to stabilize temperatures negating the need for energy intensive cooling and reducing the space heating requirements.

Sustainability and procurement

Sustainable procurement practice will be discussed under two themes or headings:

- Life cycle assessment (LCA) and
- Whole life costing (WLC)

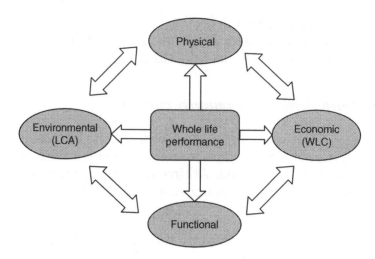

Figure 3.4 Whole life performance.

LCA describes the methods used to measure and evaluate the environmental burdens associated with a product system or activity, by describing and assessing the energy and materials used and environmental impacts released to the environment over the life cycle, from the cradle to the grave (Figure 3.4).

LCA and WLC have developed separately in response to economic and environmental issues but the two tools have much in common.

According to the Building Research Establishment (BRE) the key similarities of LCA and WLC are that both utilize data on:

- quantities of materials used,
- the service life of the materials could or will be used for,
- the maintenance and operational implications of using the products,
- end of life proportions to recycle including sale value and disposal.

Where as the key differences between LCA and WLC are (Figure 3.5):

- Conventional WLC methods do not consider the process of making a product; they are concerned with the market cost. Life cycle analysis costing on the other hand considers production.

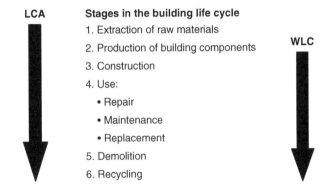

Figure 3.5 Commonality within the building cycle of issues relating to WLC and LCA (source: Edwards, Bartlett, *BRE Digest 452*).

- WLC is usually discounted to present value over time, as illustrated later in this chapter, whereas environmental impacts are not discounted.

The terms used throughout this chapter therefore will be those recommended by the BRE, as these terms are also used extensively within the construction industry, namely whole life costing and life cycle analysis. WLC can be combined with environmental or other targets, and can be used to validate a more sustainable approach, by quantifying the cost of complying with current and future environmental legislation. As such, WLC forms a critical step for organizations wishing to move towards sustainability.

Many within the construction industry relate sustainability simply to green issues, for example using only timber from sustainable sources or ensuring that a certain percentage of aggregates come from pre-used sources. However, sustainability has a much wider scope than this. Sustainable construction is the set of processes by which a profitable and competitive industry procures built assets and their immediate surroundings which:

- Enhance the quality of life and offer customer satisfaction.
- Offer flexibility and the potential to cater for user changes in the future.
- Provide and support desirable natural and social environments.

- Maximize the efficient use of resources.
- Achieve higher growth while reducing pollution and maximizing the efficient use of resources.

Some of the ways in which these goals may be achieved are by:

- Consideration of whole life costs and life cycle analysis costs to acheive procurement in line with value for money principles on the basis of asset life time costs.
- Reducing waste:
  - The application of value engineering into the procurement process can help to achieve less waste during construction and operation.
  - Each year in the UK 13 million tonnes of construction materials are delivered to site and thrown away – unused, 10% of products are wasted through oversupply, costing £2.4 billion per annum, with a further £2.4 billion being wasted in stockpiling materials. The widespread application of supply chain management techniques, discussed more fully in Chapter 4, with the application of systems such as Just In Time could drastically reduce the amount of waste.
- Reduced targets for energy consumption: Targets for energy and water consumption for new projects that meet at least current best practice for construction type and which contribute significantly to the achievement of cross-government targets agreed by the government's Green policies.
- The protection of habitat and species taking due account of the UK Biodiversity Action Plan. Biodiversity being described as: 'A recognition of the interlinked way that the planet's ecosystem works. For example, a simple hedgerow may contain hundreds of species of insects that each play their part in providing an ecosystem'. Working with the Grain of Nature: a Biodiversity Strategy for England was launched on 24 October 2002. The strategy seeks to ensure biodiversity considerations become embedded in all main sectors of public policy. The strategy sets out a series of actions for a variety of fields, including the built environment. Among other measures the strategy seeks to encourage business to

act for biodiversity in the boardroom, through the supply chain and in their management systems.

- Respect for people – this is the social part of sustainability. The well known definition based on the Bruntland Report and now adopted by the UK government is sustainability and is about the sustainable development of humans taking into account economic, social and environmental impacts. An example might be ensuring your supply chain are Investors in People, that builders are part of Considerate Constructors Scheme.
- Less pollution contribution towards goals of less pollution, better environmental management and improved health and safety on construction sites.
- Better environmental management.

Sustainability: drivers and incentives

However laudable the idea of introducing sustainability into the construction process and despite the fact that socially responsible investment in all sectors has increased by over 1000% during the latter part of the 1990s, to date there has been resistance to address the issue by clients as well as contractors. After all there has to be an incentive, by way of increased profits or sales, to adopt what has been previously demonstrated, approaches that can prove to be expensive. The comparison can be drawn with the car industry of a few years ago, when manufacturers had to convince customers that paying for safety measures and non-pollution technology was a price worth paying. First, consider the legal obligations placed on the industry.

National plans and EU programmes for sustainable initiatives

All EU member states and accession countries have been encouraged to draw up and publish plans and programmes for sustainable construction. To date the following states have done this:

- Finland
- The Netherlands

- Sweden
- The United Kingdom.

There is to date no European Directive directly relating to sustainable issues in construction although there is an agenda to issue in the near future a Commission Communication on the interpretation of the Community Public Procurement Law and the possibilities for integrating environmental considerations into public procurement. The basis of which is that while the basis for accepting bids should remain 'the most economically advantageous' at the same time integrating environmental aspects into the equation, see footnote to Chapter 4. Over the lifetime of the project, there is likelihood that the UK government or the EU will introduce or tighten regulatory or fiscal measures to control or internalize externalities, especially environmental ones. This outcome might have a significant effect on the viability of the project, and so should be taken into account in the initial appraisal.

Ecopoints

A UK Ecopoint is a single unit measurement of environmental impact of a particular product or process expressed in units, ecopoints. It is calculated in relation to impacts on the environment on the UK and therefore applies to the UK only. Ecopoints cover the following environmental impacts:

- Climate change
- Fossil fuel depletion
- Ozone depletion
- Freight transport
- Human toxicity to air
- Waste disposal
- Water extraction
- Acid deposition
- Ecotoxicity
- Eutrophication
- Summer smog
- Minerals extraction.

Ecopoints are derived by adding together the score on each of the above issues.

FTSE4Good Index

One of the more interesting developments in recent years has been the launch in July 2001 of the FTSE4Good Index. FTSE is jointly owned by the *Financial Times* and the Stock Exchange and the FTSE4Good is an index series for socially responsible investment designed to reflect the performance of socially responsible equities. It is a series of benchmarked and tradable indices facilitating investment in companies with good records of corporate or social responsibility. It was designed in response to market demand for a transparent global standard for corporate social responsibility as a growing number of companies realize the effect that corporate social responsibility has on their reputation and business risk management. The index works alongside the more familiar FTSE 100 and FTSE 250 indexes, but because inclusion in the index is based not just on financial performance the selection criteria. Companies in the index include a diverse cross section of industries including automotive, pharmaceuticals and oil and gas: Microsoft, AOL, Johnson & Johnson, etc.

A selection criterion covers three areas:

- Working towards environmental sustainability
- Developing positive relationships with stakeholders
- Upholding and supporting universal human rights wherever the organization does business.

Research for the FTSE4Good is undertaken by the Ethical Investment Research Service and its partners worldwide. In order to make the FTSE4Good selection criteria reflect current practice and thinking the index is reviewed regularly so there can be no sitting on laurels. All companies now (September 2002) receive environmental impact grading of high, medium and low depending on their business type. By March 2003 all medium and low impact companies must have a publicly available environmental policy. By September 2003 all medium impact companies must provide evidence of an environmental management system. Inclusion in the index is most likely to benefit companies who adopt socially responsible investment.

The index contains exclusions, these are:

- Tobacco producers
- Manufacturers of weapons systems
- Owners/operators of nuclear power stations.

In addition it is thought that companies with good social, environmental and ethical performance are widely considered to be better managed overall and therefore better placed to attract and retain investors, customers, suppliers and employees who share their values. Reporting on a survey of 200 chief executives, chairman and directors in 10 European countries, Business in the Community (June 2000) found nearly 80% agree that companies which integrate socially and environmentally responsible practices will be more competitive; and 73% accept that sustained social and environmental engagement can significantly improve probability.

Businesses are also being scrutinized much more intensely about their impacts on biodiversity by their shareholders, not least their investors, employees and local communities. Ignoring the issue may risk negative publicity, poor investment or even affect the licence to operate. A new standard in corporate responsibility is evident with the introduction of triple bottom line reporting: social, economic environmental. Many commercial banks are increasingly becoming aware of the issue of environmental liability. The regulatory regime in some countries (the USA is a good example) places a strong burden of responsibility on banks to ensure that the businesses to which they lend take reasonable steps to minimize their adverse environmental impacts. Therefore, unless the project developer incorporates environmental externalities into the project appraisal, it might prove difficult to persuade banks to provide finance.

Themes for action during the procurement process

The following courses of action should be considered by the procurement team when developing a strategy:

- *Re-use existing built assets* – consider the need for new build. Is a new building really the answer to the client's needs, or is there another strategy that could deliver a more appropriate solution and add value?
- *Design for minimum waste* – think whole life costs – involve the supply chain – specify performance requirements and think about recycled materials. Sustainability in design requires a broad and long term view of the environmental, economic and social impacts of particular decisions. Design

out waste both from the process and the life span. As well as the obvious definition waste can also include:
- the unnecessary consumption of land, or
- lower than predicted yield from the asset.

- *Consider lean construction* – see Chapter 1. Lean construction, in its entirety, is for many a complex and nebulous concept, however the ethos of lean with its focus on the following, is worthy of consideration:
  - continuous improvement
  - waste elimination
  - strong user focus
  - high quality management of projects and supply chains
  - improved communications
- *Minimize energy in construction and in use* – fully investigate the WLC and LCA implication of the materials and systems that are being procured. Draw up environment profiles of components.
- *Do not pollute*:
  - Understand the environmental impacts and have policies to manage them positively.
  - Construction can have a direct and obvious impact on the environment. Sources of pollution can include:
    o waste materials
    o emissions from vehicles
    o noise
    o releases into water, ground and atmosphere.
- *Respect for people* – consult all stakeholders during the procurement process.
- *Set targets* – use benchmarking and similar techniques discussed in Chapter 6 to monitor continuous improvement. UK construction industry KPIs are issued in June each year.

Whole life cost procurement

Until comparatively recently, most buildings have been conceived and built on the basis of very simple criteria, fitness for purpose corresponding to the lowest possible construction cost. Moreover, as will be demonstrated later in this chapter, in many countries of the world fiscal systems of taxation, including the UK, tend to favour high running and maintenance costs over low capital costs.

It will come as no surprise that a topic that has spent most of its life in the rarefied atmosphere of academia has no shortage of definitions and hybrids. Common terms used to describe the consideration of all the costs associated with a built asset throughout its life span are; costs-in-use, life cycle costs, whole life costs, through life costs, etc. The sheer number of alternative terms tends to create a great deal of confusion. According to *BRE Digest 452* Edwards, Barlett *et al.* (2000) the term that is recommended for adoption should be WLC. Earlier in this chapter LCA was discussed in connection with sustainability. WLC procurement is a technique that is primarily used as a *'decision-making tool'* (BS 3811), to facilitate the effective choice between a number of competing alternatives that differ, not only in their initial costs but also in their subsequent operational costs (BSI, 2000). There are a number of definitions for WLC, but one currently adopted is: *'the systematic consideration of all relevant costs and revenues associated with the acquisition and ownership of an asset.'* (Construction Best Practice Programme) and in addition the definition from the ISO Standard 15686-5 Buildings and Constructed Assets is *'a tool to assist in assessing the cost performance of construction work, aimed at facilitating choices where there are alternative means of achieving the client's objectives and where those alternatives differ, not only in their costs but also in their subsequent operations costs.'* Although WLC can be carried out at any stage of the project and not just during the procurement process, the potential of its greatest effectiveness is during procurement. This is mainly because almost all options are open to consideration at this time (Griffin, 1993). In addition, the ability to influence cost decreases continually as the project progresses, from 100% at project sanction to 20% or less by the time construction starts (Paulson, 1976; Fabrycky and Blanchard, 1991). Third, the decision to own or to purchase a building normally commits the user to most of the total cost of ownership and consequently there is a very slim chance to change the total cost of ownership once the building is delivered (HMSO, 1992; Khanduri *et al.*, 1993). Typically, about 75–95% of the cost of running, maintaining and repairing a building is determined during the procurement stage (Khanduri *et al.*, 1993, 1996; Mackay, 1999) (Figure 3.6).

First introduced to the UK construction industry over three decades ago by Dr P.A. Stone as costs-in-use; it is only recently, with the widespread adoption of Public Private Partnerships

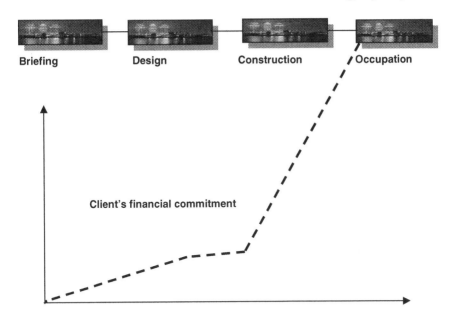

Figure 3.6 Client's financial commitment (source: Construction Best Practice Programme Fact Sheet, 1998 DETR).

(PPP) as the preferred method of procurement by the majority of public sector agencies, that the construction industry has started to see some merit in WLC. In addition, long term building owners in general are starting to demand evidence of the future costs of ownership. For example, PPP prison projects are commonly awarded to a consortium on the basis of design, build, finance and operate (DBFO), and contain the provision that, at the end of the concession period, typically 25 years, the facility is handed back to HM Prisons in a well maintained and serviceable condition. This is of course in addition to the operational and maintenance costs that will have been borne by the consortia over the contract period. A recent survey by the Building Research Establishment for the Department of Environment, Trade and the Regions (Clift and Bourke, 1999) seemed to confirm this. While further surveys by Sterner, 2000 and Clark, 2000 also indicates that building sectors in other countries have, like the UK not fully adopted whole life cost methodologies. However, for PPP consortia, given the obligations touched on above, it is clearly in the consortium's interest to give rigorous attention to costs incurred during the proposed asset's life cycle in order to mitigate operational risk. However some industry experts consider that interest in WLC by PPP consortia to be window dressing with, in reality, very little 'deep green stuff' is going on under the PPP banner. Even so PPPs have highlighted the need to consider costs other than initial capital ones.

It has been estimated that typically the relationship between capital costs, running costs and business costs in owning an office block over a 30-year period is (source: The long-term costs of owning and using buildings – Royal Academy of Engineering and Stanhope (1998)):

- Construction (capital) cost – 1
- Maintenance and operating costs – 5
- Business operating costs – 200.

The business operating costs in this equation include the salaries paid to staff, etc. Clearly in the long term this aspect is worthy of close attention by design and procurement teams.

As illustrated in Figure 3.6 the financial commitment to a built asset increases substantially once the building is complete and occupation begins. With compelling evidence like this, why have whole life cost considerations failed to be widely adopted outside of PPP? Although Stone's work was well received in academic circles, where today extensive research still continues in this field, some of which will be referred to later in this chapter, there has been and continues to be, a good deal of apathy in the UK construction industry as a whole.

One of the reasons for this lack of interest is, particularly in the private sector and the developer's market is, that during the 1980s financial institutions became less enamoured with property as an investment and turned their attention to the stock market. This move led to the emergence of the developer/trader. Often an individual, rather than a company, who proposed debt finance rather than investment financed development schemes. Where, as previously, development schemes had usually been pre-let and the investor may even have been the end user, the developer trader had as many projects in the pipeline as they could obtain finance for. The result was an almost complete disregard for whole life costs as pressure was put on the designers to pare capital costs at the expense of ultimate performance as building performance is inadequately reflected in rents and value. Fortunately, these sorts of deals have all but disappeared, with a return to the practice of pre-letting and a very different attitude to whole life costs. For if a developer trader was developing a building to sell on they would have little concern with the running costs, etc.; however, in order to pre-let a building, tenants must be certain that, particularly if they are entering a lease with a full repair and maintenance provision, that there are no 'black holes', in the form of large repair bills, waiting devour large sums of money at the end of the lease. In the present market therefore sustainability is as important to the developer as the owner/occupier. A building will have a better chance of attracting better quality tenants, throughout its life, if it has been designed using performance requirements across all asset levels, from facility (building), through system (heating and cooling system), to component (air handling unit), and even sub-component (fans or pumps).

In and around major cities today, it is clear that buildings that attracted good tenants and high rents in the 1980s and early 1990s are now tending to attract only secondary or tertiary covenants, in multiple occupancies, leading to lower rents and valuations. This is an example of how long-term funders are seeing their 25–35 year investments substantially under-performing in mid-life, thus driving the need for better whole life procured buildings.

Current whole life cost practice

Whole life cost procurement includes the consideration of the following factors:

- *Initial* or procurement costs, including design, construction or installation, purchase or leasing, fees and charges.
- *Future* cost of operation, maintenance and repairs, including: management costs such as cleaning, energy costs, etc.
- *Future* replacement costs including loss of revenue due to non-availability.
- *Future* alteration and adaptation costs including ditto.
- *Future* demolition/recycling costs.

Whole life appraisal is a personal issue (Williams, 2000) and the appraisers may include whatever they deem to be appropriate – provided they observe consistency in any cross-comparisons. The timing of the future costs associated with various alternatives must be decided and then using a number of techniques described below assess their impact. Classically, whole life cost procurement is used to determine whether the choice of say a component, with a higher initial cost than other like for like alternatives is justified by being offset by reduction of the future costs as listed above. This situation may occur in new build or refurbishment projects. In addition whole life cost procurement can be used to analyse whether in the case of an existing building a proposed change is cost effective when compared against the 'do nothing' alternative.

Again, as identified by Williams, there are three principal methods of evaluating whole life costs:

- Simple aggregation
- Annual equivalent
- Net present value.

Simple aggregation

The basis of whole life costs is that components or forms of construction that have high initial costs will, over the expected life span, prove to be cheaper and hence better value than cheaper alternatives. This method of appraisal involves adding together the costs, without discounting, of initial capital costs, operation and maintenance costs. This approach has a place in the marketing brochure and it helps to illustrate the importance of considering all the costs associated with a particular element but has little value in cost forecasting. A similarly simplistic approach is to evaluate a component on the time required to pay back the investment in a better quality product. For example, a number of energy saving devices are available for lift installations, a choice is made on the basis of which over the life cycle of the lift, say 5 or 10 years, will pay back the investment the most quickly. This last approach does have some merit, particularly in situations where the life cycle of the component is relatively short and the advances in technology and hence the introduction of a new and more efficient product is likely.

Net present value approach

The technique of discounting allows the current prices of materials to be adjusted to take account of the value of money of the life cycle of the product. Discounting is required to adjust the value of costs, or indeed, benefits which occur in different time periods so that they can be assessed at a single point in time. This technique is widely used in the public and the private sectors as well as sectors other than construction. The choice of the discount rate is critical and can be problematic as it can alter the outcome of calculation substantially. However, when faced with this problem, the two golden rules that apply are; that in the public sector follow the recommendations of the Green Book or Appraisal & Evaluation in Central Government, currently recommending a rate of 3.5%. In the private sector the rule is to select a rate that reflects the real return currently being achieved on investments. To help in understanding the discount rate, it can be considered almost as the rate of return required by the investor which includes costs, risks and lost opportunities.

The mathematical expressions used to calculate discounted present values are set out below;

$$\text{Present value (PV)} = \frac{1}{(1 + i)^n}$$

where i is the rate of interest expected or discount rate and n, the number of years.

This present value multiplier/factor is used to evaluate the present value of sums, such as replacement costs that are anticipated or planned at say 10 to 15 year intervals.

For example, consider the value of a payment of £150 that is promised to be made in 5 years time;

Assuming a discount rate of 3.5%, £150 in 5 years time would have a present worth or value of £126.30;

$$\frac{1}{(1 + i)^n}$$

$$£150 \times \frac{1}{(1.035)^n} = £150 \times 0.8420 = £126.30$$

or in other words, if £126.30 were to be invested today @ 3.5% this sum would be worth £150.00 in 5 years time, ignoring the effects of taxation.

Calculating the present value of the differences between streams of costs and benefits provides the net present value (NPV) of an option and this is used as the basis of comparison as follows.

Calculating the net present value

Two alternative road schemes have been proposed and both are expected to deliver improvements and time savings.

- *Option A* requires £10 million in initial capital expenditure to realize benefits of £2.5 million per annum for the following 4 years.
- *Option B* requires £5 million in initial capital expenditure to realize benefits of £1.5 million per annum for the following 4 years.

The significance of the results in Table 3.1 are as follows;

- *Option A* produced a negative NPV, that is to say, the costs are greater than the benefits, whereas
- *Option B* produced a positive NPV, that is to say the benefits are greater than the costs and is clearly the better

Table 3.1 Net present values

Year	0	1	2	3	4	NPV
Discount factor (PV £1)	1	0.9962	0.9335	0.9019	0.8714	
Option A Costs/ benefits (£)	−10.00 m	2.5 m	2.5 m	2.5 m	2.5 m	
Present value (£)	−10.00 m	2.42 m	2.33 m	2.25 m	2.18 m	−0.82 m
Option B Costs/ benefits (£)	−5.00 m	1.50 m	1.50 m	1.50 m	1.50 m	
Present value (£)	−5.00 m	1.45 m	1.40 m	1.35 m	1.31 m	0.51 m

alternative. A marginal or zero NPV is indicative of a do nothing option.

Alternatively, for the example given in Table 3.1, the discounting factor may be calculated as follows:

$$\frac{(1 + 0.035)^4 - 1}{0.035(1 + 0.035)^4} = 3.673$$

consequently, the NPVs for options A and B can be calculated in a single step as:

$$NPV_A = -10 + 3.673 \times 2.5 = -10 + 9.18 = -0.82 \, \text{million}$$
$$NPV_B = -5 + 3.673 \times 1.5 = -5 + 5.51 = 0.51 \, \text{million}$$

Annual equivalent approach

This approach is closely aligned to the theory of opportunity costs, that is the amount of interest lost by choosing option A or B as opposed to investing the sum at a given rate per cent, is used as a basis for comparison between alternatives. This approach also can include the provision of a sinking fund in the calculation in order that the costs of replacement are taken into account.

In using the annual equivalent approach the following equation applies:

Present value of £1 per annum (sometimes referred to by actuaries as the annuity that £1 will purchase).

This multiplier/factor is used to evaluate the present value of sums, such as running and maintenance costs that are paid on a regular annual basis.

$$\text{Present value of £1 per annum} = \frac{(1 + i)^n - 1}{i(1 + i)^n}$$

where i is the rate of interest expected or discount rate and n, the number of years.

Previously calculated figures for both multipliers are readily available for use from publications such as Parry's valuation tables, etc.

Sinking funds: a fund created for the future cost of dilapidations and renewals. Given that systems are going to wear out

Table 3.2 Whole life cost of windows

Item	Initial cost	Installation cost	Maintenance cost per day	Other maintenance costs	Life expectancy
Wooden framed window	£275	£150	£3	£100 every 3 years for preservative treatment	12 years
PVCu frame window	£340	£150	£3	None	15 years

and/or need partial replacement it is thought to be prudent to 'save for a rainy day' by investing capital in a sinking fund to meet the cost of repairs, etc.

The sinking fund allowance therefore becomes a further cost to be taken into account during the evaluation process. Whether this approach is adopted will depend on a number of features including, corporate policy, interest rates, etc.

A simple example (Table 3.3) based on the selection of window types (Table 3.2), illustrates the net present value and the annual equivalent approaches to whole life cost procurement.

This problem is a classic one, which of the two types of windows, with widely different initial and maintenance costs will deliver the best value for money over the life cycle of the building. In this example assuming a discount rate of 6% it is assumed that the windows are to be considered for installation in a Private Finance Initiative project (a barracks block for the Ministry of Defence) with an initial contract length of 25 years. Note the service life of an element, product or whole building may be viewed as:

- Technical life and Economic life – based on physical durability and reliability properties.
- Obsolescence – based on factors other than time or use patterns, for example fashion.

Table 3.3 indicates, in the right hand column, that the PVCu framed window has the lower NPV over the expected life cycle and therefore, is better value for money. However figures

Table 3.3 Whole life cost calculations

Year	Present value of £1 per annum (PV of £1 pa)	Present value (PV £1)	Initial cost (£)	Other costs (£)	Annual cost (£) (£3 × 365)	Total discounted costs = NPV of replacement + other + annual costs + initial costs (£)	Total NPV (£)	AEC (£)
*Wood framed windows**								
1	0.943	0.943	425.00		1095.00	1458.02	1458.02	1545.50
2	1.834	0.890			1095.00	974.55	2432.57	1326.81
3	2.673	0.840		100.00	1095.00	1003.35	3435.91	1285.41
4	3.465	0.792			1095.00	867.34	4303.25	1241.88
5	4.212	0.747			1095.00	818.25	5121.50	1215.83
6	4.917	0.705		100.00	1095.00	842.43	5963.93	1212.84
7	5.582	0.665			1095.00	728.24	6692.17	1198.80
8	6.210	0.627			1095.00	687.01	7379.18	1188.31
9	6.802	0.592		100.00	1095.00	707.32	8086.50	1188.90
10	7.360	0.558			1095.00	611.44	8697.94	1181.77
11	7.887	0.527			1095.00	576.83	9274.78	1175.98
12	8.384	0.497		525.00	1095.00	805.09	10079.87	1202.30
13	8.853	0.469			1095.00	513.38	10593.25	1196.61
14	9.295	0.442			1095.00	484.32	11077.56	1191.78
15	9.712	0.417		100.00	1095.00	498.63	11576.20	1191.92
16	10.106	0.394			1095.00	431.04	12007.24	1188.14
17	10.477	0.371			1095.00	406.64	12413.88	1184.84
18	10.828	0.350		100.00	1095.00	418.66	12832.54	1185.17

(continued)

Table 3.3 (continued)

Year	Present value of £1 per annum (PV of £1 pa)	Present value (PV £1)	Initial cost (£)	Other costs (£)	Annual cost (£) (£3 × 365)	Total discounted costs = NPV of replacement + other + annual costs + initial costs (£)	Total NPV (£)	AEC (£)
*Wood framed windows**								
19	11.158	0.331			1095.00	361.91	13194.46	1182.50
20	11.410	0.312			1095.00	341.43	13535.88	118012
21	11.710	0.294		100.00	1095.00	351.52	13887.40	1180.49
22	12.042	0.278			1095.00	303.87	14191.27	178.52
23	12.303	0.262			1095.00	286.67	14777.93	1178.74
24	12.550	0.247		525.00	1095.00	400.10	14878.04	1185.47
25	12.783	0.233			1095.00	255.13	15133.17	1183.82
*PVCu framed windows***								
1	0.943	0.943	490.00		1095.00	1523.02	1523.02	1614.4
2	1.834	0.890			1095.00	974.55	2497.57	1362.25
3	2.673	0.840			1095.00	919.38	3416.95	1278.31
4	3.465	0.792			1095.00	897.34	4284.29	1236.41
5	4.212	0.747			1095.00	818.25	5102.54	1211.32
6	4.917	0.705			1095.00	771.93	5874.47	1194.65
7	5.582	0.665			1095.00	728.24	6602.71	1182.76
8	6.210	0.627			1095.00	687.02	7289.72	1173.91
9	6.802	0.592			1095.00	648.13	7937.85	1167.04

Table 3.3 (continued)

Year	Present value of £1 per annum (PV of £1 pa)	Present value (PV £1)	Initial cost (£)	Other costs (£)	Annual cost (£) ($£3 \times 365$)	Total discounted costs = NPV of replacement + other + annual costs + initial costs (£)	Total NPV (£)	AEC (£)
10	7.360	0.558			1095.00	611.44	8549.30	1161.58
11	7.887	0.527			1095.00	576.83	9126.13	1157.13
12	8.384	0.497			1095.00	544.18	9670.31	1153.47
13	8.853	0.469			1095.00	513.38	10183.69	1150.35
14	9.295	0.442			1095.00	484.32	10668.00	1147.72
15	9.712	0.417		490.00	1095.00	661.37	11329.38	1166.50
16	10.106	0.394			1095.00	431.04	11760.42	1163.72
17	10.477	0.371			1095.00	406.64	12167.06	1161.28
18	10.828	0.350			1095.00	383.63	12550.69	1159.14
19	11.158	0.331			1095.00	361.91	12912.60	1157.24
20	11.470	0.312			1095.00	341.43	13254.02	1155.55
21	11.764	0.294			1095.00	322.10	13576.12	1154.03
22	12.042	0.278			1095.00	303.87	13879.99	1152.67
23	12.303	0.262			1095.00	286.67	14166.66	1151.45
24	12.550	0.247			1095.00	270.44	14237.10	1150.33
25	12.783	0.233			1095.00	255.13	14692.24	1149.36

Notes: AEC: annual equivalent cost.
* other costs are replacement costs every 12 years and preservative treatment every 3 years.
** other costs are replacement costs every 15 years.

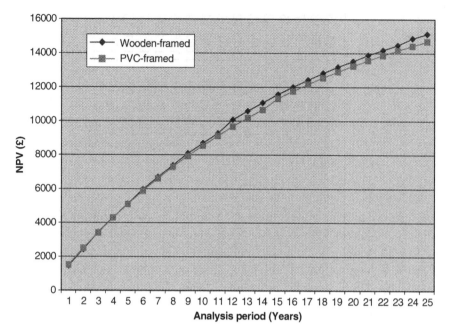

Figure 3.7 NPV of alternative window types.

should not be viewed in isolation. For example, it is known that many of the barrack's occupants, single soldiers, are heavy smokers with a tendency to stub cigarettes out on window sills, therefore wood may be the correct choice after all, as it is easier and cheaper to repair the PVCu.

Figures 3.7 and 3.8 show the net present value and the equivalent annual cost of both options. The merit of procuring PVCu windows is shown for an analysis period greater than 2.5 years, as indicated by the break-even point. To many the case using this method of reporting and presentation may not seem as convincing as first thought, as in this case, annual maintenance costs are equal for both alternatives, that is £1095.00. As differences between mutually exclusive options should be the basis of decision-making, another approach can be used. For example, Figure 3.9 shows a plot of the different costs, excluding maintenance for the two alternatives and the relative performance of the two windows is clear to see. Once again the break-even point is clearly shown.

A replacement expenditure profile, excluding cyclical maintenance and energy over a range of elements for a PFI contract over a 35-year contract period is shown in Table 3.4.

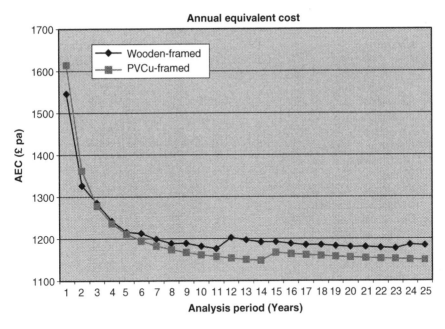

Figure 3.8 Annual equivalent costs of alternative window types.

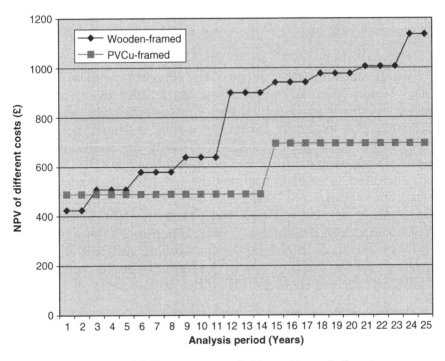

Figure 3.9 NPV of different costs of alternative window types.

Table 3.4 A replacement expenditure profile

Element	Replacement expenditure (%)
Windows/doors	22.95
Kitchens	15.79
Heating	11.82
Structural	10.63
Roofs	8.72
Bathrooms	7.79
Wiring	6.50
External areas	3.87
Internal decorations	2.48
Communal decorations	1.69
Over-cladding	1.61
Rainwater goods	1.51
External walls	0.99
Off-road parking	0.82
DPC	0.73
Security/CCTV	0.60
Door entry systems	0.51
Fire precautionary works	0.50
Porches/canopies	0.44
Plastering	0.07

Source: WLCF.

Clearly, therefore, the choice of the correct type of window or door would appear to be of critical importance to the building owner. However in reality theory and practice are often very different. For example, for many public authorities, finding budgets for construction works is usually more difficult than meeting recurring running and maintenance costs that are usually included in annual budgets as a matter of course. Many researchers (Brandon, 1987; Ashworth, 1987, 1989, 1993, 1996; Flanagan *et al.*, 1989; Ferry and Flanagan, 1991; Bull, 1993; Wilkinson, 1996; Smith *et al.*, 1998; Sterner, 2000; among others) have tried to highlight areas causing difficulties in the application of whole life costs in the industry. Kishk and Al-Hajj (1999) categorized these as difficulties on the parts of:

- the industry practices,
- the client,
- the analyst and the analysis tools currently employed in whole life costs.

What are the risks?

WLC does carry some risks. It requires a new set of skills that may be lacking in any part of the supply chain, and clients themselves are often unable adequately to describe how they expect the building or asset to be used, although construction clients are pushing for whole life costed project plans, they may be unable to interpret these correctly themselves. There is a clear requirement for considerable training in WLC across the whole of the construction industry, and particularly within the design professions.

WLC is also not an exact science, as, in addition to the difficulties inherent in future cost planning, there are larger issues at stake. It is not just a case of asking 'how much will this building cost me for the next 50 years', rather it is more difficult to know whether a particular building will be required in 50 years time at all – especially as the current business horizon for many organizations is much closer to 3 years. Also, WLC requires a different way of thinking about cash, assets and cash flow. The traditional capital cost focus has to be altered, and costs be thought of in terms of capital and revenue costs coming from the same 'pot'. Many organizations are simply not geared up for this adjustment. The common misconception that a whole life costed project will always be a project with higher capital costs does not assist this state of affairs. As building services carries a high proportion of the capital cost of most construction projects, this is of particular importance. Just as capital and revenue costs are intrinsically linked so are all the variables in the financial assessment process. Concentrate on one to the detriment of the others and you are likely to fail.

Perhaps, the most crucial reason is the difficulty in obtaining the appropriate level of information and data.

> The lack of available data to make the calculations reliable Clift and Bourke (1999) found that despite substantial amounts of research into the development of database structures to take account or performance and WLC there remains significant absence of standardization across the construction industry in terms of scope and data available. Ashworth also points out that the forecasting of building life expectancies is a fundamental prerequisite for whole life cost calculations, an operation that is fraught with

problems. While to some extent building life relies on the lives of the individual building components, this may be less critical than at first imagined, since the major structural elements, such as the substructure and the frame, usually have a life far beyond those of the replaceable elements. Clients and users will have theoretical norms of total life spans but these have often proved to be widely inaccurate in the past. The Building Maintenance Information of the Royal Institution of Chartered Surveyors was established in the 1970s. BMI have developed a Standard Form of Property Occupancy Cost Analysis, which it is claimed, allows comparisons between the cost of achieving various defined functions or maintaining defined elements. The BMI list of elements for occupancy cost analysis is listed in the Appendix. The BMI define an element for occupancy cost as; *expenditure on an item which fulfils a specific function irrespective of the use of the form of the building*. The system is dependent on practitioners submitting relevant data for the benefit of others. The increased complexity of construction means that it is far more difficult to predict the whole life cost of built assets. Moreover if the malfunction of components results in decreased yield or underperformance of the building then this is of concern to the end user/owner. There is no comprehensive risk analysis of building components available for practitioners, only a wide range of predictions of estimated life spans and notes on preventive maintenance – this is too simplistic, there is a need for costs to be tied to risk including the consequences of component failure. After all the performance of a material or component can be affected by such diverse factors as:

- Quality of initial workmanship when installed on site and subsequent maintenance.
- Maintenance regime/wear and tear. Buildings that are allowed to fall into disrepair prior to any routine maintenance being carried out will have a different life cycle profile to buildings that are regularly maintained from the outset.
- Intelligence of the design and the suitability of the material/component for its usage. There is no guarantee that the selection of so-called high quality materials will result in low life cycle costs.

Other commonly voiced criticisms of whole life cost are:

Expenditure on running costs is 100% allowable revenue expense against liability for tax and as such is very valuable. There is also a lack of taxation incentives, in the form of tax breaks, etc., for owners to install energy efficient systems. See later section on capital allowances.

In the short term and taking into account the effects of discounting the impact on future expenditure is much less significant in the development appraisal.

Another difficulty is the need to be able to forecast, a long way ahead in time, many factors such as life cycles, future operating and maintenance costs, and discount and inflation rates. WLC, by definition, deals with the future and the future is unknown. Increasingly obsolescence is being taken into account during procurement, a factor that is impossible to control since it is influenced by such things as; fashion, technological advances and innovation. An increasing challenge is to procure built assets with the flexibility to cope with changes. Thus, the treatment of uncertainty in information and data is crucial as uncertainty is endemic to WLC (Flanagan *et al.*, 1989; Bull, 1993). Another major difficulty is that the WLC technique is expensive in terms of the time required. This difficulty becomes even clearer when it is required to undertake a WLC exercise within an integrated real-time environment at the design stage of projects.

Despite the above, changes in the nature of development of other factors have emerged to convince the industry that whole life costs are important.

Whole life cost procurement – critical success factors

- Effective risk assessment – what if this alternative form of construction is used?
- Timing – begin to assess WLC as early as possible in the procurement process.
- Disposal strategy – is the asset to be owner occupied, sold or let?
- Opportunity cost – downtime.
- Maintenance strategy/frequency – does one exist?
- Suitability – matching a client's corporate or individual strategy to procurement.

Recent developments

Kishk (2001) proposed an integrated framework to handle uncertainty in WLC. It is a modified version of a previous framework outlined in Kishk and Al-Hajj (1999) and is based on the simple idea that a complex problem may be deconstructed into simpler tasks. Then, the appropriate tools are assigned a subset of tasks that match their capabilities as shown in Figure 3.1. Data is evaluated in terms of availability, tangibility and certainty. The levels of these measures increase, and hence the problem complexity decreases, from left to right. In situations where all data can be known with certainty, the problem is deterministic and can be modelled as such (Curwin and Slater, 1996). Thus, closed form solutions can provide the basis for decision-making. If outcomes are subject to uncertainty, however, alternative modelling techniques are required. According to the type of uncertainty, either the probability theory or the fuzzy set theory can be used. This way, the manner in which parameter uncertainty is described in the model can be more consistent with the basic nature of the information at hand.

The lower part of Figure 3.10 reflects the need to integrate all forms of solutions attained through various theories before

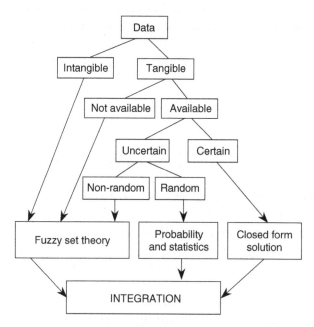

Figure 3.10 The integrated WLC framework (Kishk, 2001).

a decision can be made. Certain data, that is represented by ordinary real numbers, may be seen as special cases of fuzzy numbers (FNs) or probability density functions (PDFs), and consequently can be easily integrated with either random or non-random data as represented by FNs or PDFs, respectively. More recently, Kishk and Al-Hajj (2001a, c) have developed an algorithm to combine stochastic and subjective data as represented by PDFs and FNs, respectively, within the same model calculation. This algorithm is motivated by the fact that historic data may exist for some uncertain input parameters; and consequently, meaningful statistics can be derived for these parameters. In such cases, one might consider it more realistic to assign PDFs to these parameters. All PDFs are then properly transformed to equivalent FNs using a sound transformation technique (Kishk and Al-Hajj, 2001a). Thus, the fuzzy approach outlined in Kishk and Al-Hajj (2000a, b) can be used. These algorithms are explained and validated in the context of two example applications in Kishk *et al.* (2002).

The future of WLC

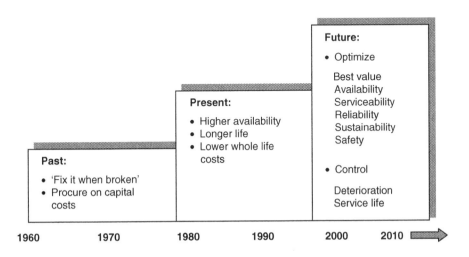

Figure 3.11 WLC – changing expectations.

The whole life cost forum

The whole life cost forum (WLCF) was formed in 1999 by a cross section of the construction industry representatives. The forum organization that is financially supported by its members who include property clients, property owners/operators, construction consultants, contractors, PFI contractors, suppliers and material manufacturers. The forum receives no financial support from government, although several departments are members. The remit was to develop a set of definitions and methodologies that could be accepted by the construction industry as a whole that would provide a degree of certainty when talking about WLC or providing quotations on a WLC basis. As the WLCF process and methodologies do not use weightings, users are able to compare options on a like for like basis and it also permits benchmarking with other projects. This is not the case, it is claimed, with other WLC methodologies. The WLCF has developed a web-based Whole Life Cost Comparator Tool which uses the WLCF methodologies and provides a secure process that does not allow users to access the calculations. Therefore any output or results from the system can be considered as authenticated, as there is no possibility of manipulation of the results.

As the finishing touches were being put to the book in early summer 2003 the comparator tool came on line. According to WLCF whatever a user's position in the construction supply chain, the tool allows secure project areas to be set up online, where suppliers can be asked to provide whole life cost estimates. The user specifies the priorities and performance parameters of the project and the user's suppliers are invited to demonstrate their optimal solutions, using detailed knowledge of their own products and services. There is a small fee for using the service and users are billed on a monthly basis. In addition to the project analysis the system also contains the capability to provide information on individual components. The author made a number of attempts to access information, on a number of routine construction elements, but without success – may be it is still early days? Attention of the author was then turned to the main system which again proved to be similarly difficult to navigate.

Effect of taxation

Capital allowances – an untapped source of added value

Tax breaks for property owners have been introduced by successive governments over a number of years, with the aim of providing an incentive for investors to replace obsolete plant and machinery with state of the art technology, or to encourage the procurement and usage of energy efficient plant and machinery. This somewhat complex area was governed by a number of pieces of legislation that were clarified and simplified in the new Capital Allowances Act (CAA) 2001. An example of the emphasis given in the new CAA is in the field of energy saving plant and machinery. The Enhanced Capital Allowances Scheme (ECAs) for energy saving plant and machinery came into effect on 1 April 2001. Under the scheme businesses are able to claim 100% first year allowances on investments in energy saving plant and machinery. The scheme supports seven energy saving technologies, combined heat and power, boilers, motors and variable speed drives, lighting, refrigeration, pipe insulation materials and thermal screens, which meet the relevant energy saving criteria included in the Energy Technology Criteria List and the Energy Technology Product List. The Finance Act 2002 added five new technologies to the list.

Capital allowances are administered by the Inland Revenue and practice is based on the CAA 2001 and well as case law. It should not be forgotten that Inspectors of Taxes enjoy a good deal of autonomy, the large amount of case law in this area is testimony to this, and therefore planning and attention to detail are essential in order to process successfully with any claim. Tax planning is a valuable tool for major construction projects and can increase the overall value of a claim by as much as 25%. By shadowing the project development during the procurement stage and recording why particular decisions are taken, tax planning compiles documentary evidence to justify the basis for claiming capital allowances.

The construction of new built assets will, for most taxpayers, be the most expensive project or type of project ever undertaken and the availability of capital allowances can often significantly reduce the post-tax cost of the project and hence add value and increase yield. Almost invariably, expenditure on a capital project will consist of some elements qualifying for a high rate of allowances and some elements qualifying for a lower rate or for no allowances at all. Almost all property owners and users, except residential, who incur expenditure on property are entitled to claim capital allowances. Most PFI projects for example are now structured to allow capital allowances to be claimed by the Special Purpose Company providing the asset.

The most common property allowances are:

- plant and machinery
- hotel allowances
- industrial building allowances

and are triggered by expenditure incurred on property under the following scenarios:

- New build, alteration and extension work incurred by tenants, leaseholders or freeholders.
- Existing property.
- Refurbishment of existing premises by landlords, owners or tenants.

Plant and machinery is the most common form of capital allowance that is available on virtually every type of commercial building and the cash value of these allowances is dependent on the product of the allowances and the claimant's rate of tax (Figure 3.12).

As illustrated in Figure 3.12 expenditure for taxation purposes falls into two categories:

- Capital expenditure with relief against liability for tax over a range of deductions between 4 and 100%, but typically at 25%, and the more valuable.
- Revenue expenditure, deducted at 100%.

Interestingly NBW Crosher and James, one of the UK's largest specialist capital allowance consultants estimate that on certain types of property that only approximately 50% of investors in property actually claim proper entitlement to fixed plant and machinery (capital allowances).

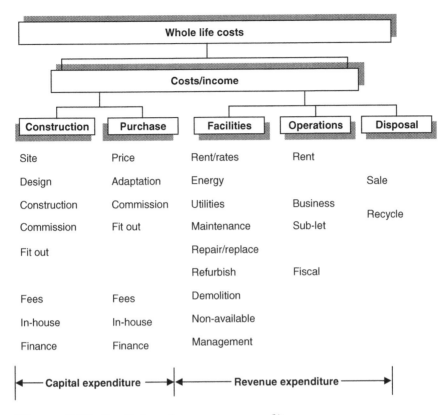

Figure 3.12 Capital and revenue expenditure.

In general a taxpayer should seek to claim relief for expenditure in the following order:

1. as a revenue deduction (100%);
2. as scientific research (traditionally very difficult to convince the Inland Revenue) or as an enterprise zone (established in 1982 and now very limited in number (both 100%);
3. as an item of plant and machinery – the most commonly claimed item (6–25%);
4. as an industrial building (4%)

Plant and machinery

Plant and machinery is the most common form of capital allowance that is available on virtually every type of commercial building. The exact definition of plant is somewhat imprecise, deliberately so, some would argue, but would typically include items such as heating and ventilating installations, lifts, fire alarms, etc. The imprecise nature of definitions in this area have led inevitably to many taxpayers testing their claims for tax relief in the courts, and many decided cases resulting from these legal battles form the basis of capital allowance practice. An example of how the proper consideration of capital allowances at the procurement stage can deliver added value in the form of tax saving is illustrated in the case; *Abbott Laboratories Ltd v Carmody (1968) 44TC 589.*

Abbott Laboratories built a total of four new buildings, three were used as industrial buildings and therefore qualified for tax relief of 4% per annum, while the fourth building was used for administration purposes. The administration block was connected to the industrial buildings by means of a covered walkway and Abbott contended that as such it was part of the industrial complex and therefore should be subject to the same relief (administration buildings do not normally qualify for industrial building allowances). The court disagreed with Abbott's claim stating that the administrative block was not sufficiently integrated into the complex to qualify for industrial building allowance. If more consideration to properly integrating the administration building into the scheme had been given at the procurement stage then this situation may not have arisen.

What are capital allowances worth?

Claims are sometimes made by consultants that in certain circumstances the value of capital allowances can be as high as 100%, more usually, depending on the specification of the building, for example with or without air conditioning, etc., the maximum values for capital allowances typically range from 2% for retail warehouse, through 20% for covered shopping malls to 40% for prestige air conditioned office blocks. Therefore, assuming that out of a total capital investment of £12 million, 33% or £4 million is identified as qualifying plant and machinery, the savings to the client would be as follows (Table 3.5):

Table 3.5 Capital allowances savings

Year	Plant and machinery (£)	Annual claim (£)	Saving per year @ 30% corporation tax per annum (£)
1	4,000,000	1,000,000	300,000
2	3,000,000	750,000	225,000
3	2,250,000	562,500	168,750
4	1,687,500	421,875	126,563
5	1,265,625	316,406	94,922
6	949,219	237,305	71,191
7	711,914	177,979	53,314
8	533,935	133,484	40,045
9	400,451	100,113	30,034
10	300,338	75,085	22,525
Total savings over 10 years			**£1,132,344**

Note that the savings can exceed the 10-year period shown above and that the annual claim is calculated on a 25% per annum reducing balance basis.

Finally this chapter will consider the question of facilities management (FM) and whether FM has becomes sufficiently mature to play a strategic role in the procurement of best value built assets?

Facilities management

FM has been defined (W.S. Atkins) as *'The creation and support of an operational environment which enhances the ability of clients to deliver and expand their core services.'* FM goals are sought to be reached, at a variety of levels, through buildings and systems maintenance to (Figure 3.13):

Another definition comes from Becker (1990) who describes FM as *'organizational effectiveness'* and Williams (2001) who describes FM as *'the process by which premises and services required to support core business activities are identified, specified, procured and delivered'*. Put more simply; although FM operates at a number of levels, it is the management of the facilities (built assets) once completed and operational. Clearly then the consideration of FM at the procurement stage should be of great importance.

- Facilitate carrying on business activity.
- Enhance perceived value for money.
- Provide a customized service.
- Deliver dependability.

Strategic level

Operational level

Figure 3.13 Strategic to operational facilities management.

First, what are 'facilities' – facilities subsumes both built assets and support services, (Williams, 2001), suggests that there is no consensus as to what activities FM should or should not embrace but here are a few important areas:

- Information technology – both hardware and software.
- Premises – operating costs, maintenance of fabric/services, cleaning/housekeeping, energy costs/waste disposal, etc.
- Support services – security, catering, storage/archives/ transport, communication systems, etc.

Much of the data that is used by facilities managers appears to rely heavily on output benchmarking, a technique described in Chapter 6. For example, in the case of an office building, a useful high level benchmark is to compare the proportion of a typical organization's total budget that is absorbed by the costs of facilities. Facilities costs are second only to staff costs and in most organizations, typically adsorb as much as 15% of revenue costs, perhaps rather less where there is a high proportion of industrial output. Each tranche of facilities, that is premises, support services and information technology should be around 5% of turnover, which is the sort of level that a commercial organization would expect its pre-tax profits to be. However, reduction of any one of these costs may not automatically result in addition to bottom line profits unless the savings can be affected without reducing performance (Figure 3.14).

Again according to Williams (2001) FM as a discipline is a latter-day phenomenon, having burst upon the business world in most of the developed countries during the 1980s and in no time at all, has achieved an extraordinary high level of recognition and status. The term 'facilities management' originated in

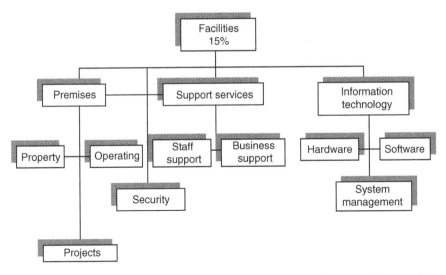

Figure 3.14 The scope of facilities management (source: Bernard Williams, 2001).

America and during the last decade or two the term has gained acceptance throughout Europe too. Bridge *et al.* (1996) see the management of operational services gaining increasing recognition as a significant factor in determining the level of corporate success achieved by a variety of organizations, whose primary business is other than the management of real property and its support services. As discussed in Chapter 4, FM plays a considerable part in the success of a project being procured using a PPP/PFI. A critical factor of FM is the policy decision concerning the source from which facilities services may be procured. Basically the choice is between:

- insourcing, that is employing labour directly within the organization, or
- outsourcing, that is to procure services from a range of external suppliers under contract to the main organization.

According to Bridge *et al.* (1996) during the early 1990s there was a trend to outsource services; however, more recently there seems to have been a swing towards insourcing. Ultimately the choice of whether to outsource or insource will be determined by the perceived benefits of each strategy. However, it is thought that while the debate continues over the source of services it is difficult for organizations to develop a more strategic approach to the procurement of facilities services. In addition, Hinks (2002) suggests that to date that there has been an emphasis on operational rather than strategic FM (see Figure 3.14) and what's more, the role of facilities management will vary according to the nature of the business environment. For example to an organization operating within a relatively stable and predictable environment, then dependability may be the top priority and emphasis and attention should be given to this criteria during the procurement stage. However, for another organization, working in a more volatile market, the ability to adapt and change quickly may be the most important criteria.

What is the process by which quality FM is delivered?

- Identify need
- Appraise options
- Produce specification
- Procurement
- Monitoring.

How is value determined in the delivery of FM?

- By added business performance
- By quantifying the potential risks to an organization for not delivering or not achieving performance levels.

Unfortunately to date most evaluation of the impact of FM decisions and policy seems to centre on quantitative databases/performance indicators over a range of products as was discussed previously in the section on whole life costs.

Throughout this book there has been heavy emphasis on the need to move away from the idea of selection by cost to selection by value for money. However it would seem that in FM at least, this move has not been made and that procurement is still very much cost based. Take for example one of the most critical FM strategic decisions, whether to procure services in house or whether to outsource. Both Barrett (1995) and Williams (2001) have reduced costs as top of the list of advantages of outsourcing. However Williams (2001) does concede that the move to outsourcing has as much to do in some cases with getting rid of intransigent managers than any real belief that outsourcing will deliver better value for money.

From a procurement perspective therefore the principal choice to be made in FM provision appears to be a simple one; is whether to insource or outsource. The principal reasons for procuring services in house or from an external source appear to be:

1. Insourcing can provide improved control through:
    – specialist knowledge and familiarity with the built asset
    – retention of staff expertise
    – continuity of standards
    – continuity of business strategy
    – risk management.
2. Cost savings through:
    – reduction in management costs
3. Expertise of resources:
    – loyalty, reliability and commitment of highly skilled staff
4. Flexibility:
    – fast response 24 hours a day, 365 days a year
5. Confidentiality and security.

On the other hand outsourcing can result in:

1. Cost savings due to few staff
2. Highly specialized knowledge not available in house
3. Flexibility in terms of location and fluctuations in demand
4. Leaves the rest of the organization to concentrate on core activities
5. Reduces risk.

Given current perceptions of FM practice it would seem as though there is still a long way to go before FM can be used at a strategic level in order to make value for money procurement decisions. At present, for a number of reasons, FM seems to be still a cost based tool.

Conclusion

The consideration, during the procurement of built assets, of the influences of the factors discussed in this chapter is no longer an option, it is a necessity. More informed clients and end users ranging from developers to PPP consortia are demanding procurement strategies that demonstrate value for money whole life cost performance. In the introduction to this chapter the question was asked – *'How many companies combine finance with management, procurement, operations and FM?'* The answer is, look around, more and more clients such as; BAA, NHS Estates, Asda, etc., already have procurement strategies that are an integral part of their corporate plan and not simply a hotch potch ad hoc approach that sits outside of the mainstream business.

Bibliography

Ashworth A. (1987). Life cycle costing – can it really work in practice? In Brandon P.S. (Ed.). *Building Cost Modelling and Computers*. E. & F.N. Spon, London.

Ashworth A. (1989). Life cycle costing: a practice tool. *Cost Engineering*, 31(3), 8–11.

Ashworth A. (1993). How life cycle costing could have improved existing costing. In Bull J.W. (Ed.). *Life Cycle Costing for Construction*. Blackie Academic & Professional, Glasgow, UK.

Brandon P.S. (1987). Life cycle appraisal – further considerations. In Spedding A. (Ed.). *Building Maintenance Economics and Management*. E. & F.N. Spon, London, 153–62.

BSI (2000) BS ISO 15686-1:2000, Buildings and constructed assets. Service life planning. General principles. British Standard Institution.

Bull J.W. (1993). The way ahead for life cycle costing in the construction industry. In Bull J.W. (Ed.). *Life Cycle Costing for Construction*. Blackie Academic & Professional, Glasgow, UK.

Clift M. and Bourke K. (1999). *Study on Whole Life Costing*, BRE Report 367, CRC.

Curwin J. and Slater R. (1996). *Quantitative Methods for Business Decisions*, 4th Edition. International Thomson Business Press, London.

Fabrycky W.J. and Blanchard B.S. (1991). *Life-Cycle Cost and Economic Analysis*. Prentice-Hall, Inc., NJ, USA.

Ferry D.J.O. and Flanagan R. (1991). *Life Cycle Costing – A Radical Approach*, CIRIA Report 122, London.

Flanagan R., Norman G., Meadows J. and Robinson G. (1989). *Life Cycle Costing – Theory and Practice*. BSP Professional Books.

Griffin J.J. (1993). Life cycle cost analysis: a decision aid. In Bull J.W. (Ed.). *Life Cycle Costing for Construction*. Blackie Academic & Professional, Glasgow, UK.

Harvey J. (2001). *Urban Land Economics*, 5th Edition. Macmillan Press Ltd.

HMSO (1992). *Life Cycle Costing*. HM Treasury, Her Majesty's Stationery Office, London.

Khanduri A.C., Bedard C. and Alkass S. (1993). Life cycle costing of office buildings at the preliminary design stage. *Developments in Civil and Construction Engineering Computing*, August, Edinburgh, ISBN: 0948749172, 1–8.

Khanduri A.C., Bedard C. and Alkass S. (1996). Assessing office building life cycle costs at preliminary design stage. *Structural Engineering Review* 8(2/3), 105–14.

Kishk M., Al-Hajj A., Pollock R., Aouad A., Bakis N. and Sun M. (2003). Whole-life costing in construction – a state of the art review. *The RICS Research Paper Series*, Vol. 4, No. 18.

Kishk M. and Al-Hajj A. (1999). An integrated framework for life cycle costing in buildings. *Proceedings of COBRA 1999 – The Challenge of Change: Construction and Building for the New Millennium*, the Royal Institution of Chartered Surveyors, University of Salford, 1–2 September 1999, Vol. 2, pp. 92–101.

Kishk M. and Al-Hajj A. (2000a). A fuzzy approach to model subjectivity in life cycle costing. *Proceedings of the BF2000 National Conference of Postgraduate Research in the Built and Human Environment*, 9–10 March 2000, University of Salford, pp. 270–80.

Kishk M. and Al-Hajj A. (2000b). A fuzzy model and algorithm to handle subjectivity in life cycle costing based decision-making. *Journal of Financial Management of Property and Construction* 5(1), 93–104.

Kishk M. and Al-Hajj A. (2001). Integrating subjective and stochastic data in life cycle costing calculations. *Proceedings of the First International Postgraduate Research Conference in the Built and Human Environment*, University of Salford, 15–16 March 2001, pp. 329–45.

Kishk M., Al-Hajj A. and Pollock R. (2002). Handling uncertain information in whole-life costing: a comparative study. *Risk Management: An International Journal* 4(3), 59–70.

Larsson N. and Clark J. (2000). Incremental costs within the design process for energy efficient buildings. *Building Research & Information* 28(5/6), 413–18.

MacKay S. (1999). Building for Life. *The Building Economist*, 4–9.

Paulson B.C. (1976). Designing to reduce construction costs. *Journal of Construction Division*, ASCE 102, 587–92.

Smith D., Berkhout F., Howes R. and Johnson E. (1998). *Adoption by Industry of Life Cycle Approaches – Its Implications for Industry Competitiveness*. Kogan Page Ltd., London, ISBN 0-749427841.

Sterner E. (2000). Life-cycle costing and its use in the Swedish building sector. *Building Research & Information* 28(5/6), 387–93.

Stone P.A. (1968). *Building Design Evaluation: Costs-in-Use.* E. & F.N. Spon.

Wilkinson S. (1996). Barriers to LCC use in the New Zealand construction industry. *Proceedings of the 7th International Symposium on Economic Management of Innovation, Productivity and Quality in Construction*, Zagreb, pp. 447–56.

Williams B. (2001). *EU Facilities Economics*. Building Economics Bureau.

Web sites

www.wlcf.org.uk

4

Procurement – systems compared

*Every time material is handled something is
added to its cost, yet not to its value*
—Henry Royce, 1907

Generally

Previously, the various drivers for change in procurement culture in the UK construction industry were discussed. This chapter will now continue to describe and evaluate the various approaches to construction procurement that have emerged recently with the promise of delivering added value for clients and the establishment of stable long-term supply chains for the construction industry.

Given a blank sheet of paper and tasked with the establishment of a procurement system from scratch, what would the essential ingredients be? Certainly the following criteria should include high on the list:

Fair, accountable and inclusive systems with

transparent procedures that are

simple to engage with for all parties from large organizations to SMEs.

For organizations or entities wishing to start, as it were, with a clean slate, the United Nations Commission on International Trade Law (UNCITRAL) Model Law provides a good starting point.

The United Nations Commission on International Trade Law Model Law on procurement of goods, construction and services

UNCITRAL is a body of the United Nations General Assembly established to promote the harmonization and unification of trade law. Its delegates and observers can and have come from many countries; as well as many ministries and professional backgrounds. This Model Procurement Law wrestles with most of the practical issues together with the problems that the UNCITRAL delegates brought to the table. *The Model Law and the Guide to Enactment*, which first appeared in 1993, sets out many methods and procedures from the most strict competitive bidding to much looser negotiating methods, the use of some of which are, in fact, not encouraged by the *Law and Guide to Enactment*. The objectives of the Model Law include maximizing competition, according fair treatment to bidding suppliers and contractors, enhancing transparency and objectivity, fostering economy and efficiency as well as curbing procurement abuses and obtaining fair value. The Model Law clearly endorses awards on the strict basis of the lowest price (actual or evaluated according to a mathematical formula), submitted by what is referred to as a responsive, qualified bidder for construction works. The methods of procurement included in the model are strikingly traditional, being:

- Single Stage Competitive Tendering.
- Two Stage Competitive Tendering.
- Restrictive Tendering.
- Single Source Procurement.

With its traditional approach, the UNCITRAL Model's main use has been by countries seeking to establish procedures for public procurement form scratch. For example, the UNCITRAL Model Law heavily influenced Poland's new procurement law.

Procurement systems compared

The remainder of this chapter as well as Chapter 5 will concentrate on new approaches to construction procurement. To avoid the criticism that this book has a pre-occupation with the so-called new procurement systems and is throwing the baby out with the bath water, the more traditional procurement strategies are discussed under *'Traditions with a twist'* towards the end of Chapter 5.

Supply chain management

Supply chain management (SCM) has been introduced previously in Chapter 1, in connection with lean construction, there now follows a more detailed analysis of SCM and SCM techniques, together with an analysis of the potential for its application to construction procurement.

Effective SCM has helped numerous industry sectors to improve their competitiveness in an increasingly global marketplace. Among the benefits that will be discussed later in this chapter, SCM has been demonstrated to provide:

- the opportunity to be innovative and learn from others;
- a true realization of the levels of performance that can be achieved;
- an opportunity to reduce waste, in all its forms.

Importantly, experience from industries other than construction have proved that, to gain the maximum advantage, SCM should be a continual process and be applied to the entire business process. Like lean production, SCM has its origins in the Keiretsu and the Kanban processes of Japan. However, unlike lean approaches, SCM is regarded by many as more easily transferred to the construction process. Even so, the popular perception of the effectiveness of SCM in construction is as illustrated in Figure 4.1, that is to say, an approach with a high applicability to production and/or manufacturing (e.g. cars and aeroplanes), but with application to construction limited to repetitive high volume buildings, as characterized by the production techniques adopted to construct fast food outlets and airport terminals!

Supply chain management – definition

From *'what is in it for me?'* to *'how can we maximize the common good?'*

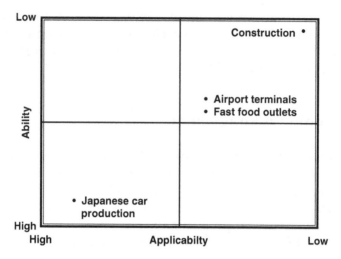

Figure 4.1 Applicability of SCM to construction.

One of the most popular misconceptions within construction networks, concerning SCM, is that supply chains are assembled for specific projects and that members of the supply chain are selected/included on promises of continuity of work flow from a prime contractor, in return for which, the supply chain members give guarantees to reduce prices and deliver to time. A perception it must be said demonstrated by some of the major headline so-called proponents of SCM. In fact this approach has little to do with SCM and more to do with the serial system of procurement popular in the 1960/1970s. Although the serial system type approach may result in an overall reduction in the cost, it does little to promote understanding by the supply chain members of the whole process, in which suppliers are just a link in the chain. Serial contracting also has few incentives to improve perform-ance and has as its main emphasis low prices instead of value and continuous performance. A supply chain can be said to be:

an association of customers and suppliers who, working together, yet in their own best interests, buy, convert, distribute and sell goods and services among themselves, resulting in the creation of a specific end product.

(National Academics, 2000)

It may be helpful to think of participants of a supply chain as the divisions of a large, vertically integrated company, bound together only by trust, shared objectives and contracts entered into on a voluntary basis. Unlike captive suppliers, that is to say, divisions of a large company that typically serve primarily, the parent company, independent suppliers are often faced with the conflicting demands of multiple customers. The integration process requires the disciplined application of management skills, processes and technologies to couple key functions and capabilities of the chain to take advantage of the available business opportunities and these will be discussed later in the chapter. SCM goals typically include higher profits and reduced risks for all participants. As players in the UK construction industry know all too well, traditional unmanaged supply chains are characterized by adversarial relationships, win–lose negotiations, short-term focus, with a primary interest on cost, with little interaction between supplier tiers and the prime contractor and with limited communications.

Arguably the UK construction industry has utilized supply chain techniques for years through the system where *ad hoc* supply chains of sub-contractors are assembled for a particular contract; the chain being disassembled at the end of the project. This *ad hoc* supply chain structure however, has limited value, but nevertheless has worked with varying degrees of success in construction due to the somewhat unpredictable nature of the construction process. However, the benefits of traditional SCM as practised in the construction industry are well below the benefits achievable from long-term partnership in the supply chain.

Procurement using a supply chain

As worldwide competition increased during the 1980s and 1990s, company profitability came under severe pressure. Having substantially reduced internal costs, industrial sectors began searching for additional opportunities for increasing their competitiveness. The drive for added value led to organizations increasingly outsourcing elements of their production. As companies analysed the amount, costs and types of value added in their own organization and compared them to capabilities available from outside suppliers, it became apparent that many elements of value could be purchased or outsourced more cost effectively. By the early 1990s re-engineering, downsizing and outsourcing became common business practices. In between the 1950s and the 1990s the percentage of sub-contracting or out-sourcing in the UK construction industry increased rapidly. The trend towards outsourcing was reinforced by the financial markets during the 1990s as analysts focused on return or yield on assets as a measure of valuing companies.

Most supply chains are organized into tiers, as illustrated in Figure 4.2, and as stated earlier should be built outside of specific construction projects.

In construction, the contracts/partnerships that create the virtual company illustrated in Figure 4.2 traditionally last for the duration of the project, with new, *ad hoc* teams being formed and disbanded, as needed, for each new job. Similar examples can be found in the defence industry, where prime contractors create *ad hoc* teams, bringing together only skills required to win and execute a specific contract. Sub-contractors, in turn have their own suppliers, who are also part of the chain. However, the *ad hoc* supply chain has limited value. It tends to work best in industries in which jobs or business opportunities are episodic

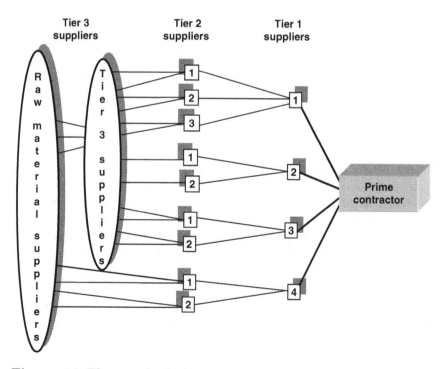

Figure 4.2 The supply chain.

and somewhat unpredictable, rather than continuous; the required capability or skill mix varies from job to job and finally where the costs of retaining a full spectrum of skills cannot be justified. For most industries however, building long-term relationships based on trust and a high level of integration yields greater benefits. Even in defence and construction benefits can often be maximized through the nurturing of long-term relationship, even if the skill set is only used on a contract-by-contract basis. How a prime contractor selects contractors and suppliers is very much down to the individual organization. For example, in Chapter 5 National Health Service (NHS) ProCure 21 is discussed, together with some of the selection criteria used by NHS Estates to select its framework partners. It is of course vital to the success of a supply chain that all its members sing from the same hymn sheet and have a similar ethos towards the delivery of value for money and quality. Also remember that the composition of a supply chain can play a vital role in the selection of a prime contractor.

As illustrated in Figure 4.2 the classic supply chain is composed of suppliers arranged into tiers or clusters, responsible to the prime contractor, therefore virtually every supplier, big or small, in the construction industry is a member of a supply chain and many suppliers participate in the supply chains of more than one prime contractor. The development and management of an efficient supply chain is an evolutionary process and the focus at strategic level should be on supplier relationships, forecasting and business objectives while equally important importance should be placed on quality, reliability, responsiveness and total cost. A supply chain organized along the lines as shown in Figure 4.2 will work well enough and deliver substantial benefits to the members; however, to be truly efficient a supply chain should, if appropriate, work towards integrating the supply chain to form supply chain communities with common goals and objectives.

Integrating the supply chain

An integrated supply chain can be defined as:

> *an association of customers and suppliers who work together to optimize their collective performance in the creation, distribution and support of an end product. The objective of integration is to*

focus and coordinate the relevant resources of each participant on the needs of the supply chain and to optimize the overall performance of the chain.

(National Academics, 2000)

The trends driving supply chain integration are:

- increased cost competitiveness,
- shorter product life cycles,
- faster product cycles,
- globalization and customization of products,
- higher quality.

The supply chain applied to construction

Where construction is concerned the supply chain for the delivery of a product clearly includes:

- the main contractor;
- other general or specialist contractors that the main contractor may employ to assist in carrying out the works;
- suppliers of materials and products to be incorporated in the works;
- suppliers of professional services such as architect, consulting engineers, etc.

Where manufacturing industries are concerned the supply chain stops at the point of delivery to a customer who will have a choice between various alternative but similar products. By comparison, in construction it is usually the client who is the ultimate consumer of the produce and has a large influence on the design and form of the finished product. Therefore, in construction there is a growing body of opinion that regards the client as an integral part of the supply chain. Interestingly, the long history of mistrust between main contractors and the sub-contractors in the construction industry has left many small sub-contractors looking to the client to defend their interests within the supply chain.

All supply chains are integrated to some extent. One objective of increasing integration is, focusing and co-ordinating the relevant resources of each participant on the needs of the supply chain to optimize performance of the whole. The integration requires the application of management techniques and technologies in order to enable the supply chain to react in a coherent manner to changes in the business environment. All supply chain members should be aware of the operational factors affecting other supply chain members.

In the production and manufacturing sectors maximum supply chain benefits are generally secured by the reduction of inventories, that is to say reducing the time that it takes for products to move through the supply chain, as typified by the success of the Dell Computer Corporation during the past decade when they reduced the time taken for components to pass through their production system from weeks to days, with future targets being set for real-time deliveries with the formation of a virtual company.

The costs, complexities and risks of fully integrating and managing an integrated supply chain can be considerable, for example:

- Managing and training costs.
- Effort devoted to becoming a better customer – see NHS ProCure 21 in Chapter 5.
- Investment in management software – see supply chain manager – later in this chapter.
- Highly integrated and interdependent supply chains become increasingly vulnerable to the risk of disruption.

Potential benefits of supply chain integration are:

- Fewer barriers and less waste of resources that do not add value.
- Increased functional and procedural synergy between supply chain members.
- Faster response to changing markets.
- Lower operational costs.
- Increased competitiveness and profitability.

Supply chain management and the construction process

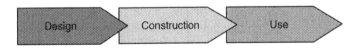

How and to what extent can SCM be applied to construction? There now follows a step-by-step approach to applying SCM to a construction project.

1. Design stage

As early as practical during the design stage, the potential of SCM should be assessed as follows:

- Carry out analysis of project.
- Identify supply chain opportunities.
- Prepare strategic sourcing programme.
- Select cross functional programme management teams.
- Identify and prioritize opportunities and issues.

2. Prepare organizations to engage in SCM

If the organization has no previous experience in SCM then workshops should be organized in order to ensure that the required levels of commitment are generated. As with partnering (see Chapter 5) if an organization is unfamiliar with SCM techniques then time must be taken to win over hearts and minds. Experiences from industry among organizations that have introduced SCM show that:

- SCM is not a quick fix and that it takes time, sometimes years, to be effectively implemented.
- Prepare for resistance, particularly from an industry like construction, renowned for its conservative approach to new initiatives. Ensure that personnel have ownership of the process – see Foreword.
- Lots of work and investment are involved in setting up the process.

To achieve the benefits of SCM a company must first determine its current and future capability, technology and capacity needs, map them against its current capabilities and then assess whether the resulting gaps can be best filled through internal development, acquisitions or outside suppliers – see later section in this chapter on capability mapping.

Having done this the next step is to perform initial supplier screening.

3. Perform initial supplier screening, meet with suppliers

Company needs should be mapped against the capabilities of potential suppliers. Suppliers should be carefully selected because the company's commitment, in many cases, will be to a long-term, intimate business relationship. The metrics for initial screening involve:

- developing cost models, with ring fenced margins (i.e. profit and offsite overheads);
- supplier selection and development of tier structure;
- define performance measurement criteria;
- define client satisfaction and finally;
- what factors add to performance, value, cost.

4. Supplier selection

In most situations suppliers are chosen on a combination of the following:

- A track record of demonstrated cost competitiveness and on-time delivery.
- Possession of proprietary capabilities.
- Demonstrated management capabilities.
- In-depth quality performance.
- Willingness to develop seamless processes and eliminate waste.
- Compatible cultures.
- Financial strength and profitability.
- Competitive technology.
- Ability to innovate.

The initial pool of potential suppliers will be reduced significantly after the initial selection process and following these, unsuccessful suppliers should be informed why they were unsuccessful. The remaining candidates should then be subjected to in-depth risk assessments to identify strengths, weaknesses and deficiencies – see also Chapters 5 and 6.

5. *Proceeding with selected suppliers*

Having selected potential suppliers the process now moves ahead by interrogating the selected supplier's costs paying particular attention to:

- base cost/profit/cost/overheads methodology;
- developing agreements;
- holding implementation workshops;
- developing joint implementation plans;
- developing systems to measure performance – see Chapter 6.

6. *Sustaining relationships with suppliers*

One of the principal ways to sustain relationships in the supply chain is through the promotion of collaborative working and transparency. The transition from traditional approaches to a collaborative approach is illustrated in Figure 4.3. During the transition emphasis should be given to:

- establishing collaborative relationships with supply chain partners and

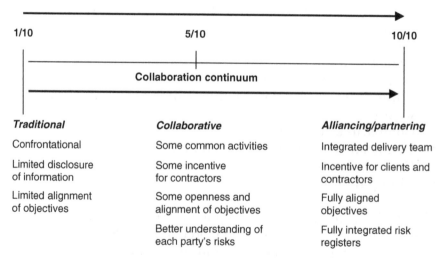

Figure 4.3 Collaboration continuum.

- developing methods to incentivize suppliers to improve their performance, such as pain/gain agreements discussed in Chapter 5.

In practice 10/10 is probably not possible, except perhaps in certain areas of the mature North Sea oil and gas industry. Realistically a degree of collaboration between 5 and 7 on the continuum is the most that can normally be achieved. As with all new procurement strategies, monitoring progress and seeking continuous improvement is a basic requirement for all supply chain members.

7. Monitor and review progress and improvement

There are two common forms of measurement – *compliance measurement* (did all parties carry out their contractual obligations?) and *performance measurement* (focuses on measuring performance against targets, time, cost, etc.) – see Chapter 6. Targets can include:

- Time
- Cost
- Change
- Safety
- Collaborative working
- Quality
- Resources.

If appropriate it may be commercially advantageous to expand the relationship into business integration of certain suppliers.

A new type of professional is required by the construction industry to facilitate the smooth operation of the supply chain – a professional who combines a variety of skills that will be described below.

The supply chain manager in construction

SCM is the management, integration and co-ordination of the whole of the supply chain, from design through to delivery of the complete built asset in order to:

- create a commercially safe environment where the effectiveness of the supply chain as a whole is more important than effectiveness of each individual company;
- deliver projects on time, to budget and to required quality;
- eliminate anything that does not add value or manage risk (waste).

A prime contractor's or principal supply chain partner's competitive advantage is highly dependent on the integrated management of the supply chain. SCM makes use of a growing body of tools, techniques and skills for co-ordinating and optimizing key processes, functions and relationships, both with the prime contractor and among its suppliers and customers to enable and capture opportunities for synergy. Supply chains must be flexible and adaptable to change in order to survive. A supply chain manager actively manages the supply chain for the prime contractor, a role well established in manufaturing sectors, but not yet commonplace in construction. Among the attributes that a supply chain manager can bring to the table is the ability to add value and manage risk. In addition supply chain managers should be market and financially aware. The supply chain manager's role can be said to operate at two levels:

- At strategic level this covers
  - set up the systems and all that goes with them,
  - relationship management,
  - knowledge sharing,
  - high level incentivization (supra pain and gain),
  - high level performance measures,
  - high level targets,
  - high level continuous improvement.
- At contract level this covers
  - commodity purchases,
  - contract incentivization,
  - contract performance measures,
  - contract targets,
  - contract continuous improvement.

The range of tools/skills commonly used by supply chain managers include the following:

- Team dynamics
- Capability mapping
- Decision matrices
- Legal frameworks
- Market analysis
- Spend analysis
- Process mapping
- Negotiation.

Team dynamics

Teamwork depends on a united approach to achieving performance goals. For SCM to be successful the members of the supply chain have to be deadly serious about implementing SCM and make a commitment to making it work. It requires a hearts and mind campaign linked to training and learning and the assembling of cross functional management teams. The team approach must also be extended in order to sustain relationships and deliver continuous improvement.

Capability mapping

Capability mapping is a technique that can be used by supply chain participants to lay out in an organized way, all of the critical functions, processes and capabilities required to design, procure and build the end product, in a construction context, a built asset. The technique involves mapping key requirements and then superimposing them on one supply chain map (see Figure 4.4). Maps of requirements can be systematically overlaid with the actual capabilities of each participant to identify gaps (e.g. deficiencies in capabilities and capacity in the supply chain). These potential problem areas then become candidates for more careful data gathering, monitoring and remedial action. Obviously not all functions can be mapped in this way, however, some functions, processes and links are more important than others and it is therefore crucial to identify the

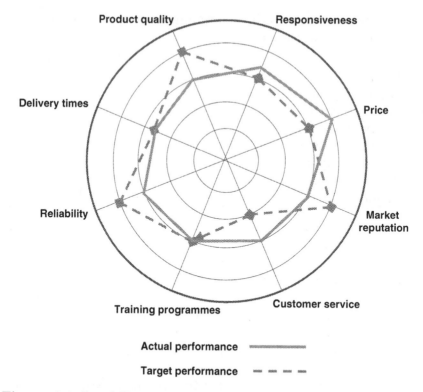

Figure 4.4 Capability mapping.

most significant ones and prioritize the allocation of resources accordingly. The following are examples of capability mapping parameters:

- Demonstrate financial strength commensurate with the risk involved in becoming part of the supply chain.
- Ability to provide consistently high quality products and services throughout its own enterprise and throughout its own supply chain.
- Compliance with accepted industry standards for quality assessment, quality maintenance and manufacturing performance (e.g. ISO 9000).
- Consistent time delivery.
- Ability to react to short lead-in times without degrading quality.
- Competitive costs.

- Emphasis on employee learning and a well-implemented training programme.
- Ability to design and produce prototypes quickly and accurately.
- Ability to provide special insight into uses for their products and methods for adapting them to non-standard applications.
- Product planning, design and development capabilities, including focus on creating and updating the supplier's own line of production.

One of the proprietary systems available in the energy sector to supply chain managers for capability mapping is First Point Assessment (FPAL). FPAL is an independent and unbiased oil and gas industry body. FPAL capability mapping is an in-depth assessment conducted from responses provided to a capability profiling questionnaire and once completed removes the need for repetitive and generic pre-qualification activity.

The assessment is carried out across a maximum of eleven business characteristics or elements with the results being expressed as a profile display, as illustrated in Figure 4.4. There is no pass, fail or overall single score as a result of the assessment and the system is designed to represent a company's capabilities and to allow a client or prime contractor to decide for themselves an acceptable standard for the work being considered.

Typical benefits claimed for the use of such a system are:

- For suppliers
  - To understand your client's assessment criteria.
  - To identify the extent that appropriate controls and systems exist in your own organization that are relevant to the industry.
  - To highlight business areas where improvement action and review may be required.
  - Used in benchmarking to provide a focus impetus for improvement actions.
  - As part of an internal communication tool.
  - As part of your marketing efforts.
  - A cost effective way advising customers about a supplier's products and services.
  - A facility to benchmark company performance.

- For purchasers
  - To access the information whenever placing work.
  - To eliminate doing their own costly and repetitive pre-qualification exercises and only do job specific pre-qualification.
  - To use the scores in a search to highlight suppliers' strengths and weaknesses and to identify where they may require to give additional support.
  - To filter out those companies who are not suitable for a particular job.
  - To provide purchasers with a structured way of identifying and selecting potential suppliers and contractors.
  - To provide objective performance information on purchasers' actual experiences.
  - To provide objective assessments of suppliers' capability thereby eliminating non-job specific pre-qualification.
  - To provide a mechanism for understanding and recording performance of both parties.
  - To provide a facility to benchmark a company's performance against competitors and identify areas for improvement.

How the system works

Both suppliers and purchasers pay a fee, for suppliers around £500 per annum and for purchasers on a sliding scale of £5000–70,000. Suppliers register their details and then FPAL evaluates the supplier across the eleven business characteristics which is then available to the supplier to verify. Once accepted the information is made available to registered purchasers. The next tool in the supply chain tool box is decision matrices.

Decision matrices

Decision matrices can take a variety of formats. They are a method to help sort through various aspects of decision in order to come to the best, most appropriate, overall choice. They help to think about choices, one criterion at a time and then combine these judgements as illustrated in Table 4.1.

The criteria considered to be essential for a particular item are entered into the matrix along the top of the table, shown

Table 4.1 Decision matrix

Criteria Alternatives	A	B	C	D	E	Total
	0.4	0.1	0.2	0.15	0.15	1.00
A	90 / 36	80 / 8	70 / 14	80 / 12	50 / 7.5	**77.5**
B						
C						
D						
E						
F						

here as A–E. These are weighted according to their perceived importance. The total of all the criteria weighting should be 1.0. The choices that are available to meet the criteria are listed under the heading of alternatives shown here as A–F. Each of these alternatives is then assigned a utility factor ranging between 0 and 100 or 0.1 and 1.0 from least acceptable (0) or best practical value (1.0 or 100). The utility value and the weighing factors are then multiplied together for all the alternatives. The alternative with the highest score is thought to be the most appropriate choice.

Legal frameworks

If procurement is taking place in the public sector then it is likely that the procedures and strategies adopted must conform to the European Union (EU) Directives on Public Procurement. Procurement, particularly in the EU, has to be undertaken in strict accordance with the appropriate legal frameworks, some of which are discussed in Chapter 6. The supply chain manager must be aware of the restrictions that are placed on suppliers, frameworks, Public Private Partnerships (PPP) and prime contractors alike with regard to EU Legislation as the tendency, recently demonstrated in the so-called Harmon Case, is to take action in the UK courts and impose heavy financial penalties for transgression.

Market analysis

Michael E. Porter's work, Competitive Strategy, Techniques for Analysing Industries and Competitors (1980) is the standard work in the field. In this classic work Porter presents his five forces, as they have become known and generic strategies. Porter identified five competitive forces that shape every single industry and market. These forces help to analyse everything from the intensity of competition to the profitability and attractiveness of an industry. The supply chain manager seeks to develop a competitive advantage over rival firms using this model. Figure 4.5 indicates the relationship and the different competitive forces, namely:

1. Barriers to entry

These are the important structural components within an industry to limit or prohibit the entrance of new competitors. The major components are: *economies of scale* (advantage of experience, learning and volume), *differentiation* (brand image and loyalty), *capital requirements* (new entrants will face a risk premium), *switching cost* involved by the customer, *access to distribution channels* and *cost disadvantages* (patents, location, subsidies).

2. Rivalry among existing competitors

In most industries, especially when there are only a few major competitors, competition will match very closely the offering of others. Aggressiveness will depend mainly on factors like number of competitors, industry growth, high fixed costs, lack of differentiation, capacity augmented in large increments, diversity in type of competitors and strategic importance of the business unit.

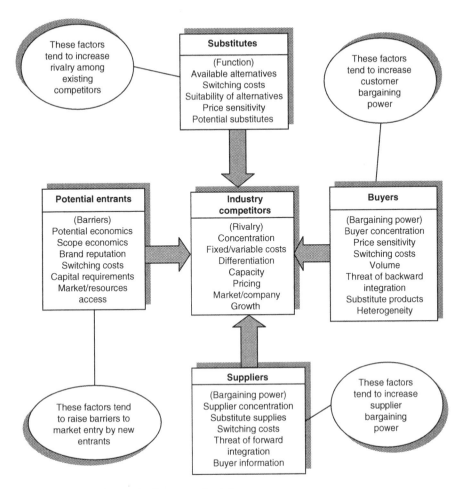

Figure 4.5 Porter's five market forces.

3. Substitutes

These are products or solutions that basically perform the same function but are often based on different technology. Depending on the level of abstraction nearly everything can be substituted. In general the only factor that really matters is a shift in technology.

4. The power of customers

Through their bargaining power buyers can force the competitors to lower their prices or force higher quality or better service. The major factors which determine the bargaining power are: volume (relative to seller sales), does the product represent a major fraction of the buyer's costs, switching costs, importance to the quality of the final product.

5. *The power of suppliers*

Suppliers can exert their bargaining power over participants by threatening to raise prices or reduce the quality. A supplier group is powerful if they are more concentrated than the industry that they sell to, or if the customer group is not import-ant for the suppliers, or if the product is not important for the suppliers.

Another technique available to the construction supply chain manager is spend analysis.

Spend analysis

Supply spend analysis (SSA) packages are available from a number of specialist providers and are used to reconcile the total spending profile of a prime contractor. SSA identifies the areas in a supply management process that can have the biggest impact on profitability. The analysis concentrates on answering the following questions:

- What is spent?
- With whom is it spent?
- How is it spent?

This sort of analysis can identify:

- Duplicate suppliers.
- If two suppliers are providing the same product or service.
- Where there is too much dependence and consequently risk on one supplier.
- Which suppliers are dependent on your business and which suppliers are critical to the well-being of the prime contractor. This sort of information may be useful in negotiations.

For example a spend analysis could be used for analysing the connection between the spending profile of a prime contractor and the number of suppliers that are used in certain critical areas. It follows that the prime contractor may be vulnerable in areas where there is apparent dependence on a few or even a single suppler, it may be prudent to consider appointing other suppliers for this service.

Process mapping

The process level has been found to be the least understood and hence the least managed level of business enterprise. While the organizational structure of a firm and the inputs and outputs are generally well documented and monitored, the interactive processes by which work is actually carried out are often largely left uncharted (Rummier and Brache, 1995). Process mapping is a tool by which these processes can be understood, documented and improved. Process mapping is complementary to benchmarking (see Chapter 6), and provides a concrete framework within which benchmarks can be set. Process mapping provides a horizontal perspective on the organization. It focuses on:

> *the value chain, which starts with the receipt of a request from a company stakeholder and terminates when that request has been answered to the satisfaction of all. Each value chain is made up of tasks that are linked across the organization. Orders are not completed by one functional silo, but rather by the cooperative efforts of individuals in different departments, functions and perhaps divisions.*
>
> (Leibfreid and McNair 1994).

At its most basic level, process mapping tools define business processes using graphic symbols or objects, with individual process activities depicted as a series of boxes and arrows (see Figure 4.6). The key stages to process mapping (Hunt, 1996) are defined as follows:

- Design a process map to relate to both things and activities.
- Distinguish what functions a system should perform from how the system is built to accomplish those functions (organizational structure).

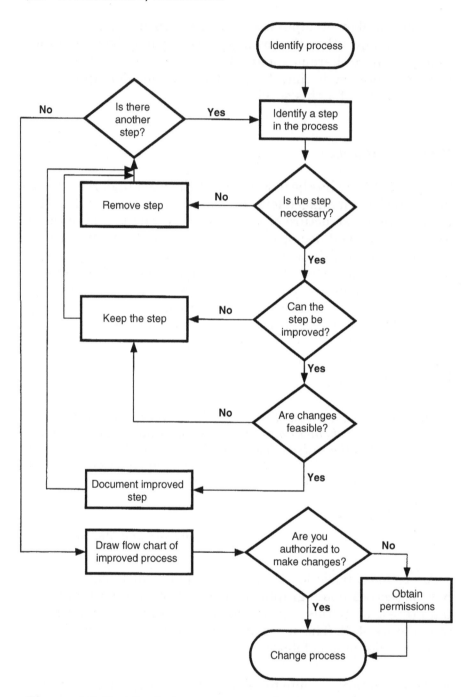

Figure 4.6 Supply chain process mapping.

- Structure the process map as a hierarchy with major functions at the top and successive process map levels revealing well-bounded details.
- Establish an informal process map review cycle to 'proofread the developing map and record all decisions in writing'.

Once the processes have been mapped, the next task is to apply appropriate measures to those processes. A flow chart, as shown in Figure 4.6, is often used for process mapping; it depicts the nature and flow in a process. Among the benefits of using process mapping are that they:

- Promote understanding of a process by explaining the steps pictorially. People may have differing ideas about how a process works. A flow chart can help gain agreement about the sequence of steps.
- Provide a tool for training employees, because of the way they visually lay out the sequence of process steps.
- Identify problem areas and opportunities for process improvement. Once a process is broken down into steps, problem areas become more visible. It is also easier to stop opportunities for simplifying and refining a process by analysing decision points, redundant steps and rework loops.
- Depict customer–supplier relationships, helping the process workers to understand who their customers are and how they may sometimes act as suppliers and sometimes as customers in relation to other people.

More specifically points in a process can be identified as high or low risk, which parts of the process add value and to what extent and what are the supply chain drivers.

The last, but by no means the least, SCM technique to be considered is negotiation.

Negotiation

The dictionary definition of negotiation is *'to talk with others to achieve an agreement.'* There is no doubt that interpersonal skills rate very highly on the list of success factors for supply chain managers and the ability to negotiate with supply chain

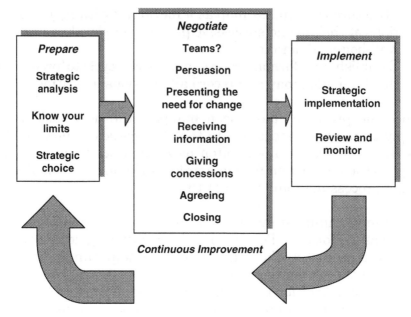

Figure 4.7 Negotiation continuum.

members and others is vital. Figure 4.7 illustrates the application of negotiating skills to the process.

Finally a footnote: experience on introducing SCM from industry and practice is that it is not a quick fix – it takes time – years. In addition be prepared for resistance and lots of work in the setting-up process and it is not a perfect system, the best contractors may find themselves on several 'A lists' and become overloaded with all the consequences that this brings.

Supply chain management in action

A main contractor is selected by a client to design and construct a major new facility with an estimated cost of £30 million with a completion time of 19 months. In order to be confident that the targets can be met it is essential that during the selection period the main contractor identifies key subcontractors/suppliers and includes them in the procurement process as early as possible.

The main contractor maintains a database of preferred subcontractors and suppliers and selection takes place on the basis of questionnaires and pre-contract meetings. The performance of the sub-contractors and suppliers has been monitored over recent contracts and the database updated accordingly.

The main contractor has taken a decision to reduce the number of sub-contractors/suppliers and to work more collaboratively with a smaller number in an attempt to reduce costs and risks and to develop a greater understanding of the various organizations cost drivers. This was done by identifying the relatively small number of sub-contractors/suppliers with the maximum spend. Having completed this process the main contractor increases its commitment to a small number of trusted and able sub-contractors and suppliers.

Sub-contractors and suppliers are selected on the basis of previous performance, ability, experience, local knowledge, resource availability and price. The main contractor actively seeks sub-contractors and suppliers who are enthusiastic about supply chain partnering. The main contractor's approach to sub-contractor/supplier supply chain partners is based on:

- Ensuring that all sub-contractors/suppliers are treated fairly, paid promptly, and integrated into the team instead of being left on the sidelines.
- Involving key sub-contractors/suppliers at an early stage and utilizing their technical expertise, experience and ideas.
- Encouraging sub-contractors/suppliers to advise of any potential problems at the earliest stage and suggest possible solutions.
- Supply key sub-contractors/suppliers with repeat business over an extended period.

Based on previous experience, the main contractor selects the key supply chain partners (sub-contractors/suppliers) and has decided to involve them in the tender process. This ensures maximized design input and establishes joint programmes that are achievable. All the partners provided tenders based on the main contractor's usual procedures. In addition the sub-contractors/suppliers are invited to attend tender briefings to ensure that they fully understood the extent of the works and to confirm the financial arrangements and to ensure that they can adequately resource the project. If supply chain partners felt unable to give these assurances then they could drop out and be replaced.

What are the advantages for the client?

As illustrated in Figure 4.8 the traditional non-integrated approach to the delivery of built assets results in price uncertainty, little opportunity for innovation and lack of client focus.

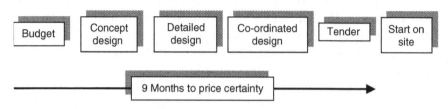

Figure 4.8 Nine months to price certainty.

Traditional approach

An integrated approach to the design and construction of built assets, unlike the traditional approach is able to take advantage of concurrent working, re-usable designs and shared experiences. For the client this enables the prime contractor to respond to clients' business needs with an earlier understanding of the required outcomes and objectives, less time and cost spent in changes leaving more time for refining the design/process and innovation – in other words a seamless process with earlier price certainty.

Other consequences

Many of the new approaches to procurement require pre-qualification of contractors and suppliers, in order to minimize the risk to the client. The Civil Engineering Contractor Association (CECA) raised a number of issues regarding this practice including:

- The use of pre-qualification databases such as the troubled public sector Constructionline.
- The costs of contractor entry into these databases – reckoned to be between £400 and £500 per year per database, on top of

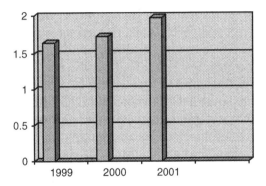

Figure 4.9 Contractor's costs of tendering as % of contract value (source: CECA).

a joining fee. It should be pointed out Constructionline's fees are much cheaper than this.
- Reduced tender periods which often proved insufficient.
- Rising costs of tendering associated mainly with the spread of design and construct contracting and frameworks and with the introduction of the quality bid.

CECA members when questioned suggested that there had been a relatively steady increase in tender costs from around 1.65% of contract value in 1999 to 2% in 2001 (Figure 4.9).

Public Private Partnerships

The Public Sector Comparator Test for Value for Money

> *After a decade of PFI, the approach should have reached maturity rather than appearing to be stuck in adolescence with all the associated angst.*
>
> (Terry Neville, University of Hertfordshire).

As has been demonstrated on many occasions winning PPP/PFI work is not simply about demonstrating familiarity with the 14-stage procurement process – it goes deeper than that. It involves detailed sector knowledge (i.e. health, educa-tion), as well as knowledge of financing, risk, EU legislation and developing innovative ways to provide and deliver public services.

More than 10 years after its introduction in the UK and after more than 500 agreed PPP deals, this controversial form of procurement still refuses to stop making headlines. PPPs currently account for approximate 12% of all public sector expenditure in construction related capital spending. In the NHS in England alone, according to the Department of Health, £8.5 billion of PPP/PFI deals have either been approved or are being considered. A PPP is a term used to describe a range of practical relationships between sectors that have varying degrees of formality and differing legal or commercial foundations. In its broadest sense PPP encompasses voluntary agreements and understandings, service level agreements, outsourcing and the PFI. The payment profile for the PPP/PFI projects compared with traditional procurement can be depicted as follows:

1. Conventional public procurement (Figure 4.10a)
2. PFI procurement (Figure 4.10b)

A commonly used PFI model for construction projects is Design, Build, Finance and Operate (DBFO) in which a private consortium builds and operates a facility, which also may include the delivery of a service, on behalf of a government agency for a period typically of 30 years, in return for a guarantee payment. Other procurement models that are used are:

*   Design, Build and Finance (DBF) schemes which exclude operational services.
*   Design, Build and Operate (DBO) schemes which rely on public rather than private funding.

In addition, other models that are available are illustrated in Figure 4.11.

Generally, PPP/PFI projects are very large complex deals that involve not only the provision and running of a built asset, but also the delivery of services to the general public. The PFI approach is most suited where the service requirements can be clearly defined at the outset and are unlikely to vary much over the lifetime of the contract. It is generally accepted that £15–20 million is the threshold for break-even on PFI projects. In a previous book *New Aspect of Quantity Surveying Practice* the mechanisms of PPP/PFI are explained in detail; however, the main controversy still surrounding this

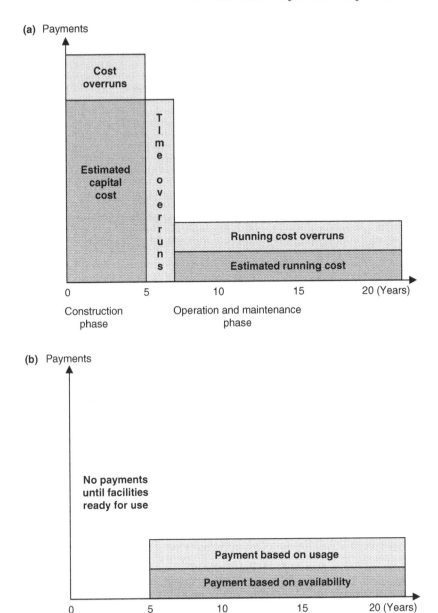

Figure 4.10 Conventional public procurement compared to PPP:
Source: Price waterhouse Coopers (2003).

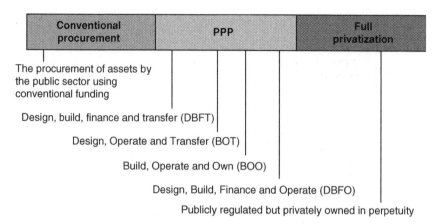

Figure 4.11 Asset procurement options (source: KPMG).

method of procurement is the ability, or otherwise, by the proponents of PPPs to produce incontrovertible and independent proof that PPP/PFI projects deliver value for money when compared with the more traditional public sector procurement strategies. In Chapter 1 some of the essential characteristics of good procurement practice were outlined and in particular the need for transparency within the process. The following spat is typical of the debate that continues about PPPs.

The Institute for Public Policy Research (IPPR) followed up its seminal report on PPPs *Building Better Partnerships* (June 2001) with a call-in for a review of PFI projects. Again, according to the IPPR there is still a need for a fully independent review of value for money in the PFI, to check both, that the schemes are expected to deliver value for money when signed and that they deliver value for predicted benefits once they are up and running. It is claimed by the IPPR that only 6% of all PPP projects have any independent examination of value for money by official audit bodies such as the National Audit Office (NAO). However, in a rebuff to the IPPR the NAO issued a statement in January 2003 dismissing the call, claiming that it did not have the resources to analyse every PFI project and that proportionately more PFI projects were reviewed by the NAO than other projects.

The debate was further fuelled when a report by Audit Scotland came to the conclusion that the methods used to assess value for money in PFI deals are 'subject to inherent uncertainty and subjectivity'. It is further claimed that substantial

Table 4.2 Savings from the PFI

	Previous experience (1999 Modernizing Construction)	*PFI experience (2002 NAO census)*
Construction projects where cost to the public sector exceeds price agreed at contract	73%	22%[1]
Construction projects delivered late to public sector	70%	24%[2]

Notes: [1] None of the increases in PFI price after contract award were due to changes led by the consortium alone. For example, in some cases the department changed some of the specifications so the price increased to reflect this.
[2] In only 8% of the PFI projects surveyed was the delay more than 2 months. No comparative data for this statistic are available for traditionally procured projects. Previous studies of traditional projects referred to the percentage of time overruns rather than the number of months.
Source: NAO (2003).

pressure is put on to public sector managers, particularly in the NHS, to ensure that the PFI option appears better than the alternatives because failure to do so will often result in a project being scrapped. Table 4.2 illustrates the predicted, often marginal, savings on PPP projects and it is this very closeness that makes the calls for a more wide ranging independent survey difficult to resist. The principal difficulties in categorically proving that the PFI provides better value than traditional procurement methods, as pointed out by one of two recent PFI reviews, *Taking the initiative – Using PFI contracts to renew council schools – June 2002, Audit Scotland*, is in the majority of cases, not easy.

PFI: construction performance

The drivers behind the many UK government initiatives to improve performance in the UK construction industry and in particular the public sector, has been that in the past, as a matter of course, projects have been delivered over budget, late and full of defects. How, therefore has the involvement of the private sector improved this lamentable record? In February 2003 the Audit Commission published its report, *'PFI: Construction Performance'*, in which it examined the construction performance of English projects procured using the PFI. The exercise was a follow-up to an earlier NAO report on Modernising Construction (1999) and comparisons were made by the NAO between the two reports. It should be noted that the 1999 Modernising Construction report did not include any PPPs, where as the 2003 study was based on a census of 38 PFI construction projects which were completed, or due to be completed by summer 2002. The hypothesis of the NAO survey was that 'The PFI will deliver price certainty ... and timely delivery of good quality assets.' And in order to investigate this the main trust of the questions asked were:

- Has there been price certainty during construction?
- Was the project completed on time?
- Is it a good quality project?

The conclusion drawn from the results by the NAO was that the census of projects generally supported the hypothesis as illustrated in Table 4.2.

On the question of quality the NAO concluded that most public sector managers surveyed were satisfied with the design and construction of their PFI buildings albeit on a very limited sample and somewhat objective.

Since the introduction of PPPs in 1992 one of the main principles is that projects procured using the PPP strategy deliver better value for money than traditional procurement paths. The model used to demonstrate value for money is the public sector comparator (PSC). The PSC is a model of risk adjusted costings of the public sector as a supplier of the outputs specified in the output specification. The PSC is comprised of a series of internal benchmarks that specify the main attributes of the entire project on a whole life cost basis, specifically:

- the standard of performance required in the delivery of outputs,
- an analysis of the cost sensitivities and financial impact of project risks.

Value for money is demonstrated when the total present value cost of private sector supply is less than the net present value of the base cost of the service, when adjusted for:

- the cost of the risks retained by the public sector,
- cost adjustments for transferable risks.

To put it crudely, to obtain approval, private sector proposals for service delivery must be less than the traditional public sector solution. However, to some observers the PSC is merely a convenient device for politicians to fall back on when claims are raised on the ability of PPP/PFI deals to deliver value for money. What is more the PSC deals with the theoretical risk transfer at the time that the contract was drawn up which in most cases is before the design was finalized, rather than an actual set of costs from comparable schemes. Further it is thought impossible to identify risks and their consequences over the 25–30-year period of a PFI deal. For example, consider the risk associated with the capability of the asset to meet the service needs, as discussed in Chapter 2, Table 2.4. A PFI concession is a long-term commitment to deliver accommodation that assists the current and future delivery of say, NHS clinical services. It is a commitment made today for the delivery of services for typically 30 years. The limited flexibility to change the scope and configuration of assets and supporting facilities procured under PFI cannot be underestimated. The impact of inflexibility will only be understood over the longer term. To gain an understanding of the types of changes that the NHS has seen over the last 10 years, beds have decreased by 25%, new outpatient episodes have increased by 27% and accident and emergency attendances increased by 23%. One of the key tests of the PFI is the ability for the private sector innovation to be transferred to the design of the new facilities, including the ability to provide long-term flexibility to cope with future changes in clinical services and facility development.

The comparison is further complicated, as very often PPP/PFI deals include land sales and/or transfer of other assets on advantageous terms to the private sector.

The PSC calculations for existing PFI deals were conducted in accordance with 'Treasury Taskforce Technical Note No. 5 – How to construct a Public Sector Comparator' as well as

'The Green Book – The Appraisal and Evaluation in Central Government' issued by the HM Treasury. Among other information, The Green Book lays down the discount rate that must be used when carrying out operations such as a PSC. For an explanation of discounting and discount rates see Chapter 3. Various commentators had for a number of years been suggesting that the discount rate in The Green Book was too high, especially in the light of sustained low UK interest rates. In April 2003 a new version of The Green Book was introduced which among other items reduced the discount rate from 6 to 3.5%. As margins between PFI deals and traditional procurement are slim in any event, this reduction could, it is felt, have dramatic effect on the outputs from the PSC. For example consider a Carlisle NHS Trust's recent PFI hospital project; using a discount rate of 6% the PSC indicated a balance of £1.7 million in favour of the PFI option; however, if the discount rate had been 3.5% instead the PSC would have indicated a balance of £12 million in favour of the traditional procurement path! The higher the discount rate the lower the net present value of payments, as illustrated in Figure 4.12.

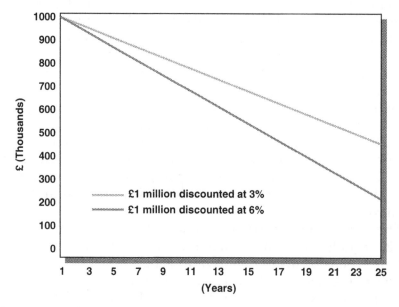

Figure 4.12 Discount rates and the PSC.

Problems with establishing the appropriate discount rate were discussed in Chapter 3 and it was established that the discount rate, which is used for evaluating projects will vary between different organization or even between projects. As investment risks are related to time, the discount rate choice is even more problematical with long-term projects that have varying risks. For example, PFI projects currently carry a higher perceived risk during the construction phase than the operation phase as once the construction phases complete the risks are changed significantly. Does this mean a different discount rate? In consequence the discount rate value is not universal in time, nor is it necessarily valid geographically on a countrywide basis. Finally the discount rate is not a fixed number, once calculated, either in time or to all phases of the project it is not a one size fits all, for example:

- Major infrastructure project may take 5 years to complete before operating for 35 years. Is it right to apply the same discount rate to both the construction and operation phases?
- During the 35-year operations contract the government may change at least seven times. How will future administrations simplify this project at the rate of return it generates at that time?
- Staff are staying in one job for shorter periods; will we still have the core competencies to run this project in 10 years?

One can appreciate from this example there are real problems with trying to develop a compound rate that covers maximum expenditure in the first 2–5 years and income for 35 years. This is one of the real problems with discounted cash flow, namely that discounting is really only effective in the early period of a long project whereas the risks change with time and after 15+ years variances in cash flow have very little effect on the initial net present value.

What factors are included for consideration when preparing a PSC? The following list gives some indication:

- Opportunity costs – the costs associated with investing capital in project A instead of project B
- Capital costs
  - Equipment acquisition
  - Buildings

- Operating costs
  - Staff costs
  - Raw materials and consumables
  - Avoidable overheads
- Sales proceeds from existing assets/land
- Asset residual value
- Relative inflation
- Third party revenues, revenues generated from operations run alongside, but not as part of core service delivery.

What is not included in a PSC?

- General price inflation
- Cost of capital
- Depreciation
- Value added tax (VAT)
- Sunk costs.

Note that the cost of capital is excluded from the calculations. The basic premise being that although private finance costs are known to be high, the transfer of risk from the public to the private sector will compensate for this.

In January 2003 the House of Commons Committee of Public Accounts published a report of its examination of the PFI deal for the redevelopment of the Ministry of Defence's (MOD) Main Building in London. The substance of the deal is that, in May 2000 the MOD signed up to a contract valued at £746 million over 30 years with the Modus consortium for the redevelopment of Main Building, MOD's London headquarters. The project is of particular interest because of the closeness of the PSC calculations. The PSC gave a central estimate for the cost of conventionally financed alternative to the PFI as £746.2 million, compared with expected PFI deal of £746.1 million. The Public Accounts Committee came to the conclusion that '*Such accuracy in long-term project costings is spurious, and the small margin in favour of the PFI deal provides no assurance that the deal will deliver value for money.*' Indeed one cannot help but agree that, a projected saving of a mere £100,000 on a project with a life span of 30 years and a total value of £0.75 billion is hardly significant or conclusive evidence that PFI in this case provides value for money (Figure 4.13). In fact it was revealed during the examination by the

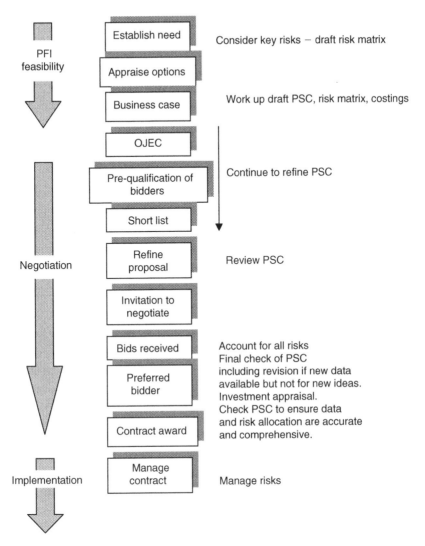

Figure 4.13 The life cycle of the PSC.

committee of the MOD project that the consortium had reduced its price by £4 million in order to undercut the PSC figure during an advanced stage of the negotiations. In this particular case the MOD used the PSC as a negotiating tool to reduce the price deal by £4 million on the day of financial close.

The House of Commons Committee on Public Accounts concluded that the two procurement routes were in fact similar in cost terms but the MOD justified the selection of Modus by

claiming that other non-quantified factors tipped the balance
in favour of the PFI deal including such considerations as:

- greater price certainty of the period of the contract than
  would be possible under alternative arrangements;
- flexibility for the MOD to reduce its occupancy levels in the
  Main Building allowing the consortia to sub-let vacant space
  and thereby reduce the amount of payments to the MOD
  from Modus;
- payments linked to services in the long term;

although quantification of the above factors has proven to be
difficult in the short term they are outside the scope of the PSC.

The analyses of the PFI compared to the notional publicly
funded alternatives (PSC) carried out by the IPPR in 2002
demonstrates that the cost advantages are narrow, see Table 4.3,
at somewhere between 0 and 9%, but typically around 5%. It is
claimed by PFI consortia that operational savings are greatest
in situations where the private sector is not only in control of
the built asset, but also the staff delivering the service, as
say, for example, in the prison service, where service providers
such as Group 4 Prison Services operate three of the current
ten privately owned prisons in the UK. At Rye Hill, Wold and
Altcourse. Altcourse, formally known as Fazakerley will be
discussed later in the chapter in connection with refinancing.
In addition the IPPR report concludes that the PSC is subject
to uncertainty and subjectivity. Current approaches to the PSC
takes no account of the higher cost of private finance compared
to council borrowing, the reason for this is given below.

Bearing in mind that the need to compare projects with different cash flows over time on a common base, it is necessary
to compare costs using discounted cash flow methodology, as
described in Chapter 3. For the PFI option, the comparison of
costs involves discounting the unitary charge proposed by the
provider, that is cost per prisoner/pupil or patient bed, over the
period of the contract. The unitary charge is itself derived from
a financial model of the provider's entire forecast cash flows
which include financing and borrowing (Figure 4.14). Calculation of the net present value of the PSC is made on a somewhat different basis, as it does not include any provision for
financing, borrowing or repayment. Instead, arising out of
the requirements to discount the PSC cash flows at 3.5%

Table 4.3 Expected PFI savings

Sector	Audit body	Date published	PSC, £ million	Value of signed PFI deals, £ million	Expected PFI savings, %
Health	NAO – West Middlesex Hospital	21/11/02	130	125	4.0
	NAO – Dartford and Gravesend Hospital	19/05/99	181.6	176.5	3.0
Schools	Audit Scotland	12/06/02			
	Falkirk		115	105	9.0
	Balfron		25	23	8.0
	Glasgow		460	434	6.0
	West Lothian		55	53	4.0
	Edinburgh		124	122	2.0
	Highland		32	32	0.0
Prisons	NAO	31/10/97			
	Bridgend		319	266	17.0
	Fazakerley		248	247	0.4
Roads	NAO	28/01/98– 09/04/99			
	A74/M74		236	200	15.0
	M1-A1		339	232	31.0
	A69		49	62	−21.0
Central depart- ments	NAO British Embassy, Berlin	30/06/00	48	38.5	21.0
	MOD – Main Building	18/01/02	748.5	746	0.3

Source: IPPR (2002).

per annum, there is an imputed cost for capital employed by the project. Imputed costs are the costs assigned to factors of production that an organization does not actually buy specifically for the project (e.g. owner's capital or capital equipment). The use of imputed cost of capital rather than the actual

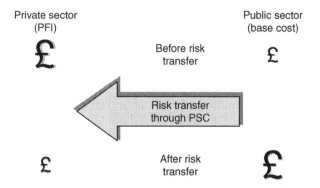

Figure 4.14 Value for money: risk transfer and the PSC.

financing costs is justified by reference to wider economic considerations, under which it can be argued that financing costs
do not have any role in public sector investment appraisal. The
level of pubic spending as a whole is a macroeconomic decision
taken by the Government taking into account judgements of
the level of borrowing appropriate to economic conditions and
the level of public capital investment. Within this quantum
choices have to be made as to which projects will go ahead.
This is the opportunity cost of capital and is thought by some
to be more relevant for the purposes of investment appraisal.
Finally, another perspective on the relevance of the PSC is that
the exercise is carried out only when the preferred bidder
has been selected and immediately prior to the signing of the
contract. In fact the decision in favour of a particular scheme
has already been earmarked on the basis of other considerations and the purpose of the PSC is to provide a cost comparison between two alternative methods of procuring the same
outcome. By their nature many of the benefits associated with
the PFI, unlike the costs, are difficult to quantify. From
the evidence of the Audit Scotland study it has not been
possible to draw overall conclusions on value for money by
comparison of the costs and benefits involved. Perhaps if PFI
deals were unbundled, that is to say if the Design, Build
Finance and Maintenance aspects were tendered for, or at
least accounted for, separately it would be easier to ascertain
whether value for money, using this procurement path has
been achieved.

The benefits that are attributed to the PFI are not unique.
Other forms of procurement have the potential to deliver

many of the benefits claimed by the PFI, however because of the insistence and pressure by central government to use only the PFI for new capital projects there has been little opportunity to develop these. The recent emphasis on PFI means that evidence of the outcomes and effectiveness of say, large-scale schools, or indeed any large capital project, procured by traditional methods is rare. The danger is that decisions in favour of PFI procurement may be driven by stereotypes or poorly performing alternatives rather than good evidence of demonstrable benefit.

In the next chapter the section on NHS ProCure 21 identifies the lack of standardization within NHS Estates resulting in a wide variety of designs and performance. If large private sector clients, such as BAA and J. Sainsbury plc have seen the benefits of standard procedures and specification, why not the public sector? The past has seen the establishment of such systems as CLASP but with only limited adoption. There is therefore a significant variation between local authorities in the cost of the schools developed. There appears to be no clear consensus on what makes for a well-designed school. There is also no systematic sharing and development of information between local authorities or even within local authorities.

The central argument advanced against the use of the PFI is that it is more expensive than public sector procurement, principally, it is said, because the public sector can borrow money more cheaply than the private sector. The UK government can borrow more cheaply as lending to the UK government is considered by lenders to be risk free, because of its capacity to raise taxes and also because the UK has never defaulted over a sovereign debt. Although in most cases the PSC was conducted professionally and rigorously (NAO) the plain fact is that if the PSC determines that the PFI option is not the best value for money, and then the project will not go ahead as no funding would be available.

It is risk transfer that in most cases proves the case for the PFI.

Further doubts were cast on the appropriateness of using the PSC to prove value for money in PFI schemes in June 2003 when the NAO published a report on the £40 million Laganside Courts Scheme in Northern Ireland. The difference between the PSC and PFI business case was only 0.4% and was only achieved by the inclusion of £7 million allowance added in to cover the costs of mothballing the existing courthouse during the currency of the PFI contract. The NAO further stated that there was little evidence of the quantification of risk transfer in the Courts scheme and in general that PSCs were subject to inherent uncertainty.

Current issues

Stepping aside from questions like 'do PFI projects increase overall levels of service?' and concentrating on procurement, the main issues surrounding the use of the PFI are related to:

1. high bidding costs;
2. refinancing and finance charges;
3. the impact of the introduction of the Competitive Dialogue Procedure by the European Commission.

1. High bidding costs

The complexity of the process makes PFI an expensive procurement method, both for the provider; in particular those bidders who fail to win the project and the public sector body. The Adam Smith Institute in a report in 1996 found average tender costs expressed as a percentage of expected total costs, across projects of all sizes, to be higher for PFI projects than for traditionally procured projects, as can be seen in Figure 4.15 below.

The report goes on to suggest that on the basis of these figures:

- PFI tendering costs are far greater than the average tender costs of other procurement methods, no matter what the project size is.
- That tendering costs are likely to be underestimated, since many of the consortia or special purpose vehicles reveal only the cost of achieving preferred bidder status. The full costs, including the contract negotiations are perhaps 1% more.
- Unlike other procurement methods, where tender costs diminish as a percentage of the total, there are no economies of scale with PFI tendering. There is a tendency for costs to increase as a percentage of the total.

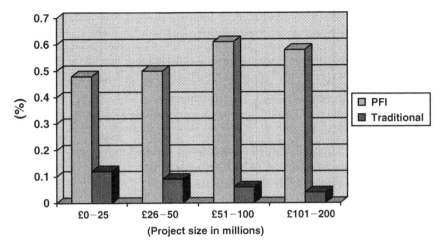

Figure 4.15 High bidding costs of the PFI.

The report finds that the total cost of tendering for a PFI project to all potential contractors to be just under 3% of expected total costs while for traditional procurement the total costs accounted for just less than 1% In July 2002 the Chief Executive of John Mowlem reported that his company was spending between £5 and 7 million a year on preparing PFI bids. Other concerns centred on the lack of consistency in the process with bids often taking between 18 months and 2 years to come to fruition. Bidding was described as a costly and unpredictable process with, for example, each education authority managing the process in its own way. In another example, by February 2003 the bid costs for the London Underground had reached a staggering £300 million of which the Tubelines Consortium had spent some £140 million. High costs must exclude parties that do not have the financial means to enter the PPP marketplace and in the long term this could be detrimental to the bidding process and could lead to lack of bidding capacity allowing main contractors to cherry-pick projects, while smaller organizations would be left out. The typical procedure for first phase of competition (Invitation to Negotiate Stage (Table 4.4), i.e. the selection of pre-qualified bidders, usually three in number, for the main PFI competition can involve the submission by prospective consortia of a whole range of information that requires careful and detailed consideration by the client. The tender documents can amount to several hundred pages, for example:

- Background and invitation to negotiate.
- An overview of the project including clear instructions on how to prepare the bid, it includes:
  - a summary of the objectives and requirement of the project by the client;
  - an overview of the procurement process;
  - the requirements for bid submission;
  - the approach to bid evaluation.
- Output specification – The output specification explains the scope of the services, including expected standards and regulations. Other matters included are: planning, minimizing disruption, etc. Design performance standards, including whole life costs and maintenance are also detailed.
- Draft project agreement – This represents the client's position on key issues, subject to negotiation with the bidders. Bidders are requested to comment on the document.

How then can bidding costs be reduced?

- Reduce the stages for tendering from the current recommended 14 stages, an option currently under review by the Office of Government Commerce.
- Reduce time up to best and final offer.
- Eliminate the best and final offer stage.
- Reduce the number of bidders to two or three.
- Develop the brief as fully as possible with improved project definition before issue to bidders.
- Reduce the need for up-front detailed design.
- Increasing and retaining public sector expertise.
- Standardization of PFI contracts.
- Do not ask bidders for full due diligence before preferred bidder stage, as this operation is often repeated on the request of the financiers before financial close, much to the frustration of the rest of the team who see matters that they thought were agreed and settled unbundled for reconsideration.

Set up costs including the cost of consultants.

Illustrative role of advisors acting for the public sector procurer only in the PFI process. PFI binding stages are give below:

	Legal advisors	Financial advisors
1. Development of the business case	Inform timetable	Options appraisal Clarify objectives Review data Inform timetable
2. Work up business case	Outline project documentation Procurement rules Relevant legislation Establish vires position of public sector	Bid strategy Indicative VFM analysis Initial view on accounting treatment Affordability review Identify costs and risk allocation
3. Tendering and negotiation	Prepare project agreement	Determine selection criteria Prepare tender documents Evaluate bids
4. Negotiations	Negotiate contractual issues Employment issues	Financial evaluation of bid Analyse funding options Negotiate with short-listed bidders Risk adjusted VFM analysis Establish accounting treatment
5. Award of contract	Inform negotiations on detail of contracts	Final VFM analysis Full business case Financial close
6. Service commencement		Performance evaluation

2. Refinancing and financing charges

Refinancing is an established technique whereby improved financial terms can be obtained in projects where risks have been demonstrated to be successfully managed; however only one of the four early PFI projects had arrangements to share refinancing gains. Alarm bells over the structure of PPP/PFI project finance were first raised over the case of the refinancing of Altcourse, formerly Fazakerley, prison in Liverpool.

Group 4 Prison Services Limited opened the prison at the end of 1997, it was one of the first PFI projects to come on stream. The financial backers were suspicious of this new form of procurement and in particular the fact that income to Group 4 Prison Services would not start to flow until the project was completed and fitted out and ready to receive inmates. Accordingly they perceived the potential for risk to be high and put a high price tag on the finance. However, shortly after it opened, in 1999, the operators were able to renegotiate their bank loans, in the light of reduced exposure to risk, with the result that returns to Group 4 increased from 16% to 39% resulting in a windfall profit of £10.7 million, of which only £1 million was given to the public sector. During the procurement process the public sector, probably through lack of experience, had not considered the benefits that such a refinancing deal would bring. In 2000 the NAO and the Committee of Public Accounts both issued reports on the Fazakerley affair with a recommendation that government departments should share in the benefits of successful PFI projects. Figure 4.16 illustrates the relationship between risk and returns in a typical PFI contract. Once the required service has been brought into operation, the project risks are lowered, as the risks associated with the commencing the service delivery are no longer relevant.

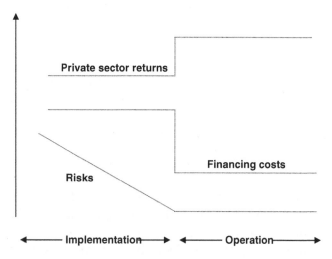

Figure 4.16 PFI refinancing update (source: NAO, 2002).

In July 2002 the OGC published its new refinancing guidance. Post Fazakerley the Office of Government Commerce carried out a large programme of work to change the approach of departments and the market in new PFI contracts and is generally seeking a 30% share of future refinancing gains on early, that is existing PFI deals and for most new contracts a 50% share of refinancing gains will be secured. What is more the private sector is required to seek approval for most refinancing situations. The benefits of the refinancing may be taken as a cash transfer to the public sector agency, or by way of a reduced unitary charge over the remaining contract period.

The sharing of refinancing gains is of course good news for the public sector agencies as well as the opponents of the PFI; however the impact that the new refinancing policy may have on the value for money of deals and the reaction by the private sector is complex and will require careful supervision and audit.

The costs of finance

One of the main arguments for using PFI procurement is that although it is more expensive for the private sector to borrow money than the public sector, to finance PFI deals, that the greater expertise of private sector management and risk management compensates for this additional charge. In its examination of the MOD Main Building PFI deal the House of Commons Committee of Public Accounts came to the conclusion that '*Although relative financing costs are crucial to decisions on PFI projects, the approach to investment appraisal normally used by departments does not enable a comparison to be readily drawn between the financing costs of a PFI deal and conventional procurement. It is not clear that current methods for PFI deals are the most efficient or the cheapest. Neither is it obvious why the improvements in risk management that are said to flow from PFI need necessarily involve expensive private financing.*'

Comparisons of the cost of borrowing by the government and the private sector are difficult to make however certain facts are incontrovertible:

- Governments borrow through the National Loans Fund (NLF), which is backed by tax revenues and so is virtually risk free and hence the cheapest way of raising funds.

- Private sector companies which have no guarantees are inherently riskier and hence borrow on less advantageous terms.

According to figures contained in the House of Commons Research Paper 01/117 based on figures in the *Financial Times* on 13 September 2001, the yield on 10 year government bonds at the close of play on 12 September 2001 was 4.9%. Similarly dated corporate bonds ranged between 6.9% for Gallaher and 5.6% for Halifax. This suggests that the extra borrowing cost of corporate bonds is at least 1.5% higher than government bonds. However, in many PFI deals the extra costs of finance have a truly significant impact on the balance sheet. For example in the case of the Skye Bridge the NAO calculated that the extra cost of finance was £4 million on a total project cost of £28 million. Similarly, the House of Commons Committee of Public Accounts in its report on the PFI deal for the redevelopment of the MOD Main Building concluded that *'Closer attention to financing costs would have been particularly helpful during the 16 months it took MOD to close the deal.'* Even in such a large PFI deal as this it would appear that all sides in the procurement process were amazingly naive when it came to finance, leaving the Committee of Public Accounts to comment that *'a cannier approach to the financing markets prior to closing the deal all might have helped secure savings on this project.'* The factors that prompted this comment was when the cost of financing the £746 million project increased by £60–185 million shortly before financial close – as is normal in PFI deals the financing cost risk remains in the public sector! This is because the exact cost of finance is not known until during due diligence in the advanced stages of financial close with the preferred bidder, at which point the public sector agencies are in the majority of cases faced with a *fait accompli* as illustrated in the above MOD case.

According to Audit Scotland finance costs range from between 8 and 10%, that is 2.5–4% higher than the public sector. This translates into a cost of between £0.2 and 0.3 million per annum for each £10 million borrowed and may be the equivalent of approximately 10% of total estimated costs over the contract period. There is therefore concern that the PFI can lead to the public sector paying large percentages of its budget to the private sector in the form of capital, which of course, adds nothing to service delivery.

Therefore, to summarize, managing the PFI procurement process remains expensive and resource hungry for both the public and private sectors due to:

- The length of the procurement process, the Audit Commission has drawn attention to the fact that the gestation period for a PPP/PFI school can be as long as 4 years from provisional central government approval to its opening.
- The current procurement system involves the use of many private sector consultants who, on a range of projects scrutinized by Audit Scotland, resulted in fees of between £0.3 and 2.4 million per project or between 5% and 15% of core construction costs.
- According to Audit Scotland even comparatively small and uncomplicated projects can result in local authority costs of £0.2.
- Private sector set-up costs are high and recovered as part of the unitary payment, that is the annual sum paid to private sector operators over the contract period, £1–9.2 million (Audit Scotland).
- PFI does leave a legacy; the certainty of costs that PFI brings removes the ability of local authorities to alter elements of their budgets.
- PSC – there is a lack of robust and credible data upon which to create a PSC. The Audit Commission – The PSC has lost the confidence of many people.

Bundling of PFI projects

Bundling is the grouping of projects or services within one project structure in a manner which enables the group to be financed as one project. The key benefits are that this allows small projects to be financed by increasing the overall debt within the bundle to an economic level and allows the various projects to cross collateralize each other. Bundling is being introduced because capital markets funding will tend to concentrate on larger projects and is therefore not available as an option for smaller projects. The transaction costs on projects with a capital value of around £10 million can be disproportionately high and severely affect return and value for money. Bundling projects can provide a reasonable return after operating and debt service costs are addressed. It can also spread between different projects and locations. Bringing projects together for financing may well require further investigation as the following issues will need to be considered: partial completion or termination of the project(s), compatibility of facilities and services, geographical constraints, different commencement times that require funding to be available at different times.

On 28 May 2003 the Department of Health outlined its latest, most radical proposal to realize the government's aspiration to procure and built 100 hospitals by 2010. In answer to the criticism that the PFI approach is too lengthy and costly, the Health Minister John Hutton announced that a consortium would bid to build and run two or three hospitals at once. The theory is that running only one competition would reduce the likelihood of delays and that bid costs could fall by as much as two-thirds.

3. Recent changes in EU Procurement Directives

In May 2002 the European Commission issued its final proposals for amendments to the Public Procurement Directives. It was the first major revision to the legislation since its first drafting 20 years or so earlier. The objectives, according to the EU were:

- modernization in order to take account of new technologies and changes in the economic environment,
- simplification to make procedures more understandable,
- flexibility in order to meet the needs of public purchasers and economic operators.

Currently in the UK the Negotiated Procedure option is used in PFI contracts because it is generally accepted that in many complex PFI contracts uncertainty exists as to precisely; how the contractor will achieve the output requirements of the public sector, how much it will cost and whether those requirements can in fact be achieved within the current marketplace. Under the Open and Restricted Procedures there is very limited scope to negotiate once tenders have been received. In contrast the Negotiated Procedure allows public authorities wide flexibility to negotiate contract terms after selection of the authority's preferred contractor. The use of the Negotiated Procedure for PFI contracts however, causes the EU some concern who considers that many PFI contracts are not sufficiently complex to justify the use of the Negotiated Procedure and that there is a danger that unfair competition may be encouraged. The European Commission therefore has introduced a new procedure, known as the Competitive Dialogue. This is a procedure *'in which any economic operator may request to participate and whereby the contracting authority conducts a dialogue with the candidates admitted to that procedure, with the aim of developing one or more several suitable alternatives to meet its requirements and on the basis of which the candidates chosen are invited to tender.'* Once the final technical specification has been settled by the public sector awarding authority it is obliged to continue the process with not less than three consortia and invite them to submit bids. Once this process has been completed the contract is then awarded without the opportunity to negotiate further. Concern exists for a number of reasons. First, it would be wasteful for the public sector to have negotiated details of proposed contracts with at least three consortia. Second, the costs to various consortium members of providing extremely detailed bids when there is less chance of success will greatly increase private sector costs. The new directive is expected to be fully operational by 2005. Many observers believe that in order to preserve the UK PFI model that many public authorities may opt for the Negotiated route and argue their decision with the Commission.

Finally if the European Commission and EU member states are determined to press ahead with the 'greening' of public procurement, as seems to be the case then according to the European Construction Industry Federation (FIEC) a number of solutions need to be considered:

1. Clients should be encouraged to assess tenders on the basis of the economically most advantageous tender, balancing price, quality and whole life costs, for which the quality assessment should include quality factors.
2. Client, architects and engineers will in future need to take much more detailed account of environmental aspects of their designs, especially life cycle analysis and whole life cost considerations.

3. Tenderers may be encouraged to put forward alternative technical solutions that take account of environmental aspects.
4. The award of contracts on the basis of concessions, that is to say where the contractor enters into a contract to construct and subsequently run the facility.

Conclusions

Its (PFI's) compelling attraction as far as the National Health Service is concerned is that we can get more hospitals built more quickly.

Alan Milburn, Secretary of State for Health
giving evidence to the House of Commons
Health Committee, June 2002.

The question facing clients therefore with PPP/PFI procurement is not so much, 'is this the most appropriate strategy for me but rather, how can the PPP/PFI procurement process be refined and streamlined to the point where value for money, and transparency are self-evident'.

Bibliography

ACCA (April 2002). *PFI: Practical Perspectives.* The Certified Accountants Education Trust.

Audit Scotland (June 2002). *Taking the Initiative – Using PFI Contracts to Renew Council Schools.*

Carr B. and Winch G. (1998). *Construction Benchmarking: An International Perspective.*

dti (2002). *Construction Statistics Annual 2002,* HMSO.

Department for Education and Skills (2003). *Building Schools for the Future.*

Hunt V.D. (1996). *Process Mapping: How to Re-engineer Your Business Processes.* Wiley, New York.

International Finance Services Ltd. (December 2002). *Public Private Partnerships – UK Expertise for International Markets 2003.* IFSL.

Lema N.M. and Price A.D.F. (1995). Benchmarking: performance improvement towards competitive advantage. *Journal of Management in Engineering.*

National Audit Office (November 2002). *PFI Refinancing Update.* HMSO.

National Audit Office (February 2003). *PFI: Construction Perform-ance.* HMSO.

National Audit Office (June 2003). *The Laganside Courts.* HMSO.

Nightingale. M (2003) Now it's critical. *Building*, 30 May, 27.

Pollock J.A. *et al*. (November 2001). *Public Services and the Private Sector.* Catalyst Trust.

Treasury Taskforce (1999). *Technical Note No 5 – How to Construct a Public Sector Comparator.* HMSO.

Web sites

www.yescombe.pwp.blueyonder.co.uk/Links.htm

www.nao.gov.uk/

www.audit-scotland.gov.uk

www.ogc.gov.uk/

www.bmj.com/

www.dti.gov.uk/

5

Procurement – relationship contracting

Introduction

This chapter will continue to review some new procurement and contractual arrangements, in current use in the UK construction industry, that aim to promote a more inclusive approach to the procurement process including prime contracting, alliancing, partnering and frameworks (including NHS ProCure 21), etc.

The age of collaboration

What do construction clients demand from the construction industry; in particular large clients who procure built assets on a regular basis? Figure 5.1 illustrates the range of attributes that informed clients are demanding from the construction industry. Top of the list of demands is the ability to innovate and provide bespoke value added solutions to construction related problems combined with the minimum exposure to risk. Consider the following typical example: a large supermarket chain has decided to earmark certain of their city and town centre stores for redevelopment. The sites chosen are mainly occupied by low rise structures and it is proposed to enhance the value of these sites by redeveloping the areas over the sales floors to provide a range of accommodation from residential to commercial. The problem for potential contractors is that the stores must be able to carry on trading as normal during the redevelopment process and contractors are tasked with the problem of how to achieve this by developing innovative approaches to carrying out the works. Second on the list is assured cost: this can be achieved by using contractors and

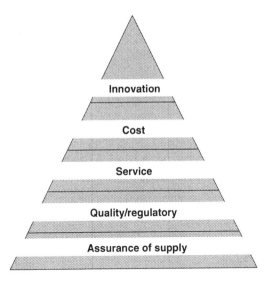

Figure 5.1 Client's demands.

suppliers operating well developed, fully integrated and efficient supply chains as well as fully accountable procurement strategies, such as partnering and alliancing. As demonstrated in other sectors efficient supply chains can also help to deliver the remaining three client demands of service, quality and assurance of supply.

Prime contracting, joint ventures, alliancing and partnering

Generally

During the 1960s and 1970s diversification was regarded by many as the gospel for business success and as a consequence many large organizations went on a buying spree that took them into technologies and markets that were to prove an expensive experience. However, during the 1990s firms began to divest themselves of their acquisitions, as success was now perceived as 'sticking to the knitting' and core business competencies. Today, increasing specialization and the growing complexity of technology has meant that companies that could once have carried out projects largely using their own resources now have to form teams of specialists to complete the task. There has developed therefore, a number of legal and contractual arrangements to facilitate the formalization of organizations pooling their resources and these can be said to be:

- Prime contractor led projects
- Joint ventures
- Partnering

and to this list can now be added Alliancing. These procurement strategies will now be examined.

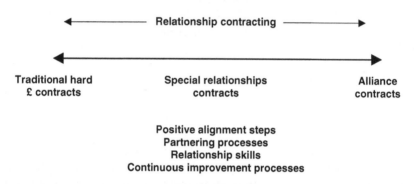

Figure 5.2 The spectrum of contract forms.

Prime contracting

A prime contractor is defined as an entity that has the complete responsibility for the delivery and in some cases, the operation of a built asset and may be either a contractor, in the generally excepted meaning of the term, or a firm of consultants. To date most prime contractors are in fact large firms of contractors, despite the concerted efforts of many agencies, such as National Health Service Estates, to emphasize the point that this role is not restricted to traditional perceptions of contracting. In some models in current use, prime contractors take responsibility not only for the technical aspects of a project during the construction phase, including design and supply chain management but also for the day to day running and management of the project once completed. This may include a contractual liability for the prime contractor to guarantee the whole life costs of a project over a pre-determined period for as much as 30 years. Favoured by some clients because when this procurement strategy is adopted there is one point of contact/responsibility for the entire project, instead of a client having to engage separately with a bewildering range of specialists. High profile clients using this approach include the Ministry of Defence in the pioneering Building Down Barriers Initiative. The approach to prime contracting differs from public private partnerships/private finance initiative (PPP/PFI). The prime contractor's obligations are usually limited to the delivery and the facilities management of the asset(s), there is no service delivery involved, as in the case of a PFI project for a school or hospital and the financial aspects of this approach are much less significant than is the case with PPP/PFI projects.

Joint ventures

Joint ventures are arrangements by which two or more firms collaborate on a project either as; a contractual joint venture, or a special joint venture company. A contractual joint venture exists where parties agree to work together, but no separate legal entity is formed. The management of the venture is overseen by a joint steering committee drawn from the firms involved. However, in a joint venture company a separate legal entity is created that trades independently from the co-operating organizations, but in which all are shareholders. The advantage of the joint venture company is that liability can be limited unlike partnerships for example where it cannot. Used commonly in the French construction industry, where the standard form of construction contract contains provision for this approach.

Partnering

Partnering, dates from the mid-1980s, especially in the UK and the USA and has been used as a generic term embracing a range of practices designed to promote greater co-operation (Barlow *et al.*, 1997). Not to be confused with partnerships, where all the partners may be liable in full for the debts of the partnership, even to the extent of their personal possessions. Partnering will be fully discussed later in the chapter.

Alliancing

The terms alliancing and partnering do not have the same legal connotations as partnership or joint venture and there has been a tendency, particularly in the construction industry, to apply them rather loosely to a whole range of situations, many of which clearly have nothing to do with the true ethos of partnering or alliances, which is to be regretted.

Alliancing and partnering agreements are not unique to the construction industry. For example, in the US alone, between 1988 and 2001 more than 100,000 alliances have been formed including companies that are thought to be resource rich and self-sufficient, like Ford and IBM. Alliance activity of the top 1000 US firms is expected to account for 35% of revenue in 2002, up from less than 2% in 1980. These approaches are divided and catagorized as follows:

- Strategic alliances can be described as two or more firms that collaborate to pursue mutually compatible goals that would be difficult to achieve alone. The firms remain independent following the formation of the alliance. Alliancing should not be confused with mergers or acquisitions.
- A project alliance is where a client forms an alliance with one or more service providers; designers, contractors, supplier, etc. for a specific project and this section will continue to concentrate on this aspect of alliancing (see Figure 5.3).

One high profile construction project success story for project alliancing in recent years was the construction of the $155 million National Museum of Australia in Canberra that was opened to the public in March 2001. The project was the first time that project alliancing had been used to procure a construction project in Australia and is widely considered to have delivered outstanding results, namely:

- Completion on time
- Completion within budget
- A very high score for quality – 8 points out of 10.

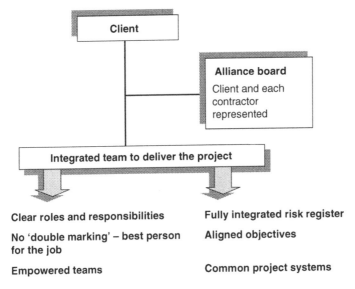

Figure 5.3 Alliancing structure.

For the National Museum project a number of alternative procurement routes were examined by a public federal government hearing during 1997. Finally project alliancing was selected for the National Museum of Australia project because there was:

- an absolute cap on funding;
- a very high expectation of quality on this Australian centenary flag ship project;
- a 'must meet' deadline.

The principal features of a project alliance are as follows:

- The project is governed by a project alliance board, that is composed from all parties to the alliance that have equal representation on the board. One outcome of this is that the client has to divulge to the other board members far more information than would, under other forms of procurement, be deemed to be prudent.
- The day-to-day management of the project is handled by an integrated project management team drawn from the expertise within the various parties on the basis of the best person for the job.
- There is a commitment to settle disputes without recourse to litigation except in the circumstance of wilful default.

- Reimbursement to the non-client parties is by way of 100% open book accounting based on:
  - (i) *100% of expenditure including project overheads*
    Each non-client participant is reimbursed the actual costs incurred on the project, including costs associated with reworks. However, reimbursement under this heading must not include any hidden contributions to corporate overheads or profit.
    All project transactions and costings are 100% open book and subject to audit.
  - (ii) *A fixed lump sum to cover corporate overheads and a fee to cover profit margin*
    This is the fee for providing services to the alliance, usually shown as a percentage based on 'business as usual.' The fee should represent the normal return for providing the particular service.
  - (iii) *Pain/gain mechanism with pre-agreed targets*
    The incentive to generate the best project results lies in the concept of reward, which is performance based. A fundamental principle of alliances is the acceptance on the part of all the members of a share of losses, should they arise, as well as a share in rewards of the project. *Risk*: Reward should be linked to project outcomes which add to or detract from, the value to the client. In practice there will be a limit to the losses that any of the alliance members, other than the client, will be willing to accept, if the project turns out badly. Unless there are good reasons to the contrary it may be expected that the alliance will take 50% of the risk and the owner/client the remaining 50%. *The sharing of pain*: Gain is generally based on objectively measurable outcomes in key performance areas, such as:

    - Time of delivery
    - Safety
    - Environmental compliance
    - Industrial and community relations.

    Performance based remuneration ensures that some of the contractor's remuneration, the profit margin referred to in point (ii), is at risk unless it achieves the indicators (see Figure 5.4).

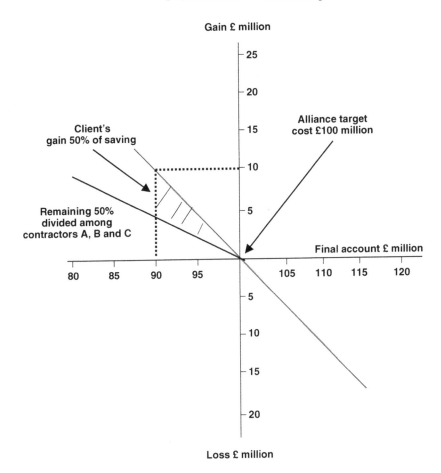

Figure 5.4 Pain and gain mechanism.
Note: Assume that the alliance has four members, who have decided to share the gain and loss percentage as follows:
Client 50%; Contractor A 20%; Contractor B 20%; Contractor C 10%.

For example, in an alliance project with a total value of £50 million, alliance member, Contractor A, decides that the maximum amount of exposure to loss that their organization is prepared to accept is 10% or £5 million, so that in the case of an over spend of £2 million then Contractor A's exposure will be limited to £200,000 or 10%. The setting of a 10% limit for loss however also limits the extent of any share of gain, therefore if the project final cost turned out to be £45 million then the share of the saving for Contractor A would also be limited to 10% or £500,000.

Therefore, alliance members form a quasi-joint venture, because they operate at one level as a single company, however they do not merge their companies in any legal sense. They remain independent but they must work with each other in order to meet the key performance indicators (KPIs) – see Chapter 6, to realize risk and reward. Therefore if the project fails to meet agreed project KPIs then all alliance members share the loss.

There are some significant legal and financial aspects that need to be put in place in an alliance agreement, but while important, it is the behaviour of the parties which determines whether an alliance will be successful. Choose your alliance members carefully!

Therefore given the operational criteria of an alliance it is vitally important that members are selected against rigorous criteria. These criteria, usually demonstrated by reference to previous projects undertaken by the prospective alliance members, vary from project to project but as a guide could be as follows.

Demonstrated ability to:

- complete the full scope of works being undertaken from the technical, financial and managerial perspectives;
- re-engineer project capital and operating costs without sacrificing quality;
- achieve outstanding quality with an outstanding track record;
- innovate and deliver outstanding design and construction outcomes;
- demonstrate safety performance;
- demonstrate conversance with sustainability issues;
- work as a member of an alliance with a commitment to non-adversarial culture and change direction quickly if required;
- have a joint view on what the risks are and how they will be managed.

Figure 5.5 is based on the procedure used to appoint alliance members for the Australian National Museum project. The risk/reward mechanisms for this project was calculated over three areas:

- *Time*: Substantial monetary penalties were agreed to all non-client alliance members if the project was completed late.

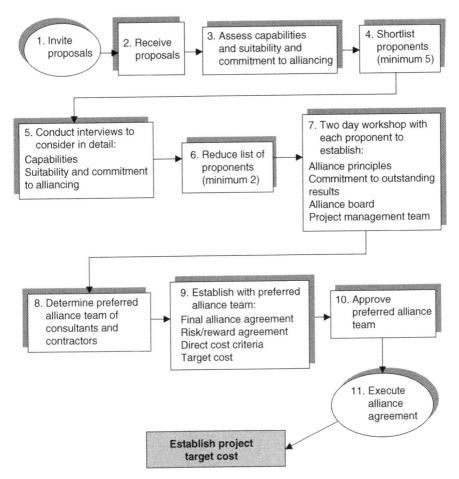

Figure 5.5 Alliancing procedure.

On the other hand there was no monetary benefit to non-client alliance members if project completed early.

- *Design integrity*: If design integrity was not maintained there was a substantial cost penalty to the non-client alliance members.

Maintenance of design integrity was verified by an independent panel that reviewed the design as it developed and documented.

The verification ensured that alliance members had to consider the impact of design changes on design integrity.

- *Quality*: The Australian government funded a quality pool to reward outstanding results in:
  - Workmanship
  - Cultural excellence
  - Safety
  - Environment
  - Public relations.

The benchmark for measuring outstanding results was 'business as usual' of the alliance parties, a score higher than business as usual generated a monetary reward and vice versa. The measurement was verified by an independent panel.

Establishing a target cost and dealing with variations

One of the key tasks on the establishment of the alliance board is the development of the target cost – it is not possible to do this before this point. The target cost is intended to be the best estimate of what the integrated teams think the project will cost. It is important that all alliance members feel comfortable with target cost which should allow for inherent uncertainties consistent with the state of knowledge at the time of preparation. Under a conventional contract the tender sum is just the starting point with subsequent variations and claims resulting in a substantially higher final account cost to the client. In a project alliance the target cost must allow for all matters that would normally be the subject of a variation under a conventional say JCT (98) contract. The alliance members collectively assume responsibility for all sorts of risks that are normally retained by the client under a traditional approach for example:

- design changes,
- late delivery by suppliers,
- inclement weather

and reasonable provision has to be made within the target cost for such items.

The circumstances under which variations arise are limited. Generally, normal changes due to design development are not considered. Changes that could give rise to variations are:

- significant increases or decreases in the scope of work, for example adding in new buildings, parts of buildings, or facilities;
- fundamental changes in the performance parameters.

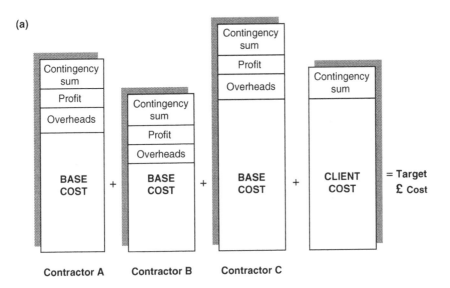

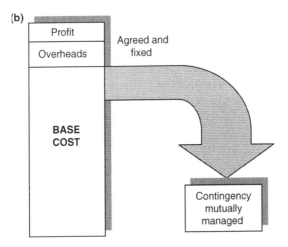

Figure 5.6 (a) Traditional and (b) alliance approach to procurement.

If the alliance members agree that a variation is declared then members' fees as well as the target cost are adjusted.

Some of the benefits of the alliance approach are illustrated in Figure 5.6. In the traditional approach to procurement each contractor, as well as the client, will include a sum for contingencies for unforeseen circumstances, whereas in an alliance profit and overheads are agreed in advance and the contingency allowance can be reduced and managed collectively.

How can clients be assured that alliancing produces better value for money?

It is difficult conclusively to demonstrate that the outcome when using alliancing will deliver cost reductions compared to a more conventional delivery approach. More likely it is the assurance of:

- Timely completion as a result of:
  - better management of changes,
  - increased levels of innovation,
  - better collective management of risks and opportunities.
- Quick, co-operative and effective response to unpredictable events.
- Emphasis on opex rather than capex.
- Potential for lower capital costs as a result of:
  - inherent buildability,
  - sharing of cost savings,
  - innovative execution strategies.

Under a traditional procurement model using competitive lump sum tendering the contractor submitting the lowest price is tested against what the market will bear. However, when using alliancing the client has no way of knowing whether the target cost is the best value for money. However, history has proved that traditional lump sum contracts have consistently failed to deliver projects to budget and to time, whereas with alliancing, a group of highly dedicated and focused people working together in a collaborative non-adversarial environment, will deliver a project to cost and to time.

Although there have been some high profile success stories with alliances, in general failure rates for alliances are high at around 50–60%, a salutary lesson for those about to dash head-long into this form of procurement. Reasons for alliance failure and conversely critical success factors have been identified as:

- poor partner selection,
- mismatch of organizational structures or culture,
- poor systems for information sharing,
- lack of trust,
- lack of commitment.

and finally, the small print.

In this culture of mutual trust and openness, is there a need for lawyers? The answer is of course – yes! Lawyers have an important role to play in ensuring that the intention of the parties is enshrined in a properly structured and legally effect-ive alliance agreement. Some of the more notable features of typical alliance agreements that set them apart from standard forms of contract include:

- Collective performance obligations
- Details of pain/gain conditions together with statement of extent of liabilities
- Details of audit arrangements for each alliance member.

Partnering

Generally received positively by the construction industry, the use of the term partnering in connection with procurement emerged in the USA in the early 1980s. The US process indus-try became concerned about the lack of competitiveness of its engineering contractors. As a result it established a client-based round table that concluded that long-term relationships with a few contractors should result in cost savings. Partnering there-fore originated as a technique for reducing the costs of com-mercial contract disputes, improving communications, process issues and relationships are the principal focus. The term part-nering rather than partnerships was chosen because of the legal connotations of the partnerships. However to some observers partnering is an ambiguous term to which at least half a dozen different perspectives may be applied.

Overall, partnering has produced mixed results in terms of improved performance. For instance, to some observers, partnering has been seen to try to impose a culture of win/win over the top of a commercial framework which remains inherently win/lose. The verbal commitments during the partnering process, even if genuine at the time, are not enough to withstand the stress imposed by gross mis-alignment of commercial interests. Some critics (Howell *et al.*, 2002), go further and describe partnering as nothing more than a programmatic 'Band Aid' on the current construction system whose fundamental weaknesses gave rise to its adoption.

There is also evidence, according to CECA, that in the civil engineering industry, it seems that some clients have entered into partnering arrangements with contractors and have let framework agreements without fully appreciating all that is required for these arrangements to be successful in terms of delivering better value for their investment in construction. In particular, some seem not to have looked much beyond the subsidiary objective of these arrangements, which is to secure savings in costs of procurement and contract administration. There is evidence of what has been labelled 'institutional pressure', that is to say clients and contractors feel that they must move in the directions in which the Latham and Egan reports are pointing them, but there is a danger that they will begin to move on the basis of insufficient knowledge and understanding of what is required.

Despite these criticisms, the widescale adoption of partnering in construction procurement has spread rapidly and has become the backbone of many worldwide government initiatives, including the UK, receiving endorsements by Latham and Egan. Previously in this chapter alliances were described and before continuing it is important to understand the differences between partnering and alliancing. The major differences between alliancing and project partnering are given in Table 5.1.

For example, in project partnering one supplier may sink or swim without necessarily affecting the business position of the other suppliers. One entity may make a profit, while the other entity makes a financial loss. However, with alliancing there is a joint rather than a shared loss, therefore if one alliance party under-performs then all the parties are at risk of losing.

In April 2002 Masons, a leading law firm specializing in construction matters, published a report entitled 'Partnering – the Industry Speaks', which contained the results of a survey of

Table 5.1 Differences between partnering and alliances

	Partnering	*Alliances*
The form of undertaking	Core group with no legal responsibilities. Binding/ non-binding charters used in 65% of partnering arrangements	Quasi-joint venture operating at one level as a single company
The selection process	Prime contractor responsible for choice of supply chain partners	Rigorous selection process
	Project can commence while selection continues	Alliance agreement not concluded until all members appointed
The management structure	By prime contractor Partnering advisor	Alliance Board
Risk and reward mechanisms	Partners' losses not shared by other members of the supply chain	Losses by one alliance member shared by other members

over 1000 participants within the construction industry, ranging from contractors, sub-contractors and consultants. Of the organizations questioned, 79% confirmed that they had been engaged in partnering, to some extent, but only 6% of organizations had more than 50% of their workload based on partnering arrangements. When questioned about the potential benefits of partnering the respondents replied:

- avoidance of disputes/blame culture,
- improved profitability,
- mutual objectives rather than mutual gain,
- greater client satisfaction.

Partnering has been described as a process whereby the parties to a traditional risk transfer form of contract, i.e. the client, the contractor and the supply chain, commit to work together with enhanced communications, in a spirit of mutual trust and respect towards the achievement of shared objectives (Figure 5.7). As with alliancing, there are two approaches, strategic and project partnering. The differences are as follows:

- Strategic partnering is concerned with a range of work, often unspecified at the time that the contract is made, over a

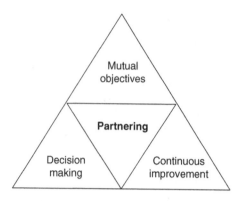

Figure 5.7 Constituents of partnering (source: dti Construction Best Practice).

period of time. The motivations are to achieve consistency and predictability of workload, to take out waste and to achieve continuous improvement through experience and learning.

- Project partnering is much more focused on a single project.

The three essential elements of partnering are that the individuals and firms involved in a construction project should agree to:

1. *Mutual objectives*

Mutual objectives to which all parties are fully committed. All the parties open up about their own objectives so that they get a better understanding of what each organization is trying to achieve. Following this, a set of mutual objectives can be drawn up, which in the case of project partnering are specific to a project taking into account the individual organizations' objectives, that can form the basis of the partnering charter. Examples of mutual objectives are illustrated in Table 5.2.

2. *Decision-making*

Second, partnering requires agreement on how decisions are made, including how any disputes will be resolved. One of the basics of partnering is empowerment and the idea that decisions

Table 5.2 Partnering objectives

Objective	How achieved
1. Improve efficiency	1. Co-operation
2. Cost reduction	2. Continuous improvement
3. Cost certainty	3. Early action on high risk areas
4. Enhanced value	4. Buildability, value engineering
5. Reasonable profits	5. Predictable progress
6. Reliable product quality	6. Quality assurance/TQM

should be made and problems solved at the lowest levels. In addition, partnering places importance on the establishment of non-contractual ways of quickly resolving such matters.

3. Continuous improvement

Finally, the parties must commit to seeking continuous and measurable improvement in their performance. KPIs are set by the partnering parties who decide among themselves what should be measured and what target should be set. Typically the items measured are:

- base cost,
- time,
- quality.

There will obviously be many road blocks in trying to introduce a system like partnering into a traditional construction organization and these can be categorized as cultural and commercial issues.

Cultural issues

Cultural realignment for an industry is difficult to achieve. The approach generally adopted by organizations is to organize partnering workshops during which the approach is explained. This includes:

- The establishment of *trust* – a vital ingredient in the partnering process.
- Emphasizing the importance of *good communications* between partnering parties.

- Explaining that many activities are carried out by *team-work* and how this is achieved in practice.
- The establishment of a *win/win* approach.
- Breaking down traditional management structures and *empowering* individuals to work in new ways.

Commercial issues

Organizations wishing to be involved in partnering arrangements should be able to demonstrate capabilities in the following:

- Establishing a *target (base) cost* that is based on open book accounting and incorporating pain/gain schemes. Profit margins should be established and ring fenced.
- A track record of using *value engineering* to produce best value solution.
- A track record of using *risk management* to identify and understand the possible impact of risk on the project.
- An ability successfully to utilize *benchmarking* to identify how to measure and improve performance.
- The willingness to take a radical look at the way they manage their business and to engage in *business process re-engineering* if necessary to eliminate waste and duplication.

So where do the potential problems and pitfalls lie, or is partnering the simple route to trouble-free contracting? The Masons' survey, *Partnering – The Industry Speaks*, highlighted the most significant potential pitfalls of partnering as:

- Having *too high expectations*. Parties will have to work hard to overcome amicably and fairly, problems and conflicts that arise.
- Having *disparate objectives*. Partnering requires the alignment of objectives and an understanding of an interest in each other's success.
- Setting *unrealistic risk allocation*. It is essential that parties have a realistic attitude to risk and how to manage it efficiently and effectively.
- *No 'buy-in' at all levels*. Full benefits in the procurement process will only be achieved if all levels of the process embrace the partnering approach.

Partnering contracts

A common sentiment expressed by various commentators is that partnering is a non-legal or 'moral' relationship that sits alongside the formal legal/contractual relationship between the parties. Some argue that we need to step away from the traditional methods of procurement and project delivery and if this means scrapping the formal written contract, so be it. However, not everyone is of this view. One school of thought is that it is important to make partnering arrangements legally binding. The contrary view being that partnering is a consensual process and that its purpose is lost if parties are forced to do it. As part of the partnering process a so-called non-binding partnering charter is sometimes used to lay over a traditional, adversarial form of contract. The charter uses language, such as honesty, trust, co-operation and shared objectives. Using this approach parties try to get partnering benefit with the re-assurance that they can return to the old ways if partnering is not successful. Funders often prefer this approach. Regardless of which view is taken, most lawyers do agree that the parties need to spell out in clear terms, whether the partnering arrangements are legally binding or not. Until relatively recently, a common approach was for the parties to contract in the usual way and then to set about formulating their partnering arrangements. Often this involved senior management having meetings to thrash out a partnering charter or agreement, which might or might not be legally binding on the parties.

In the past 4 or 5 years, specific contracts and contractual amendments have been drafted to implement partnering as part of the formal contract process. These include:

- The Association of Consultant Architects Ltd. (ACA) Standard Form of Contract for Project Partnering (PPC 2000).
- The New Engineering Contract (NEC) has option X12 set of clauses, this in effect is a Standard Form of Contract underneath a partnering agreement.

In addition the following two alternatives are available:

- No contract but a partnering charter
- Standard form of contract (JCT 98) and partnering charter.

Project Partnering Contract 2000 (PPC 2000)

Even though partnering has been described as a state of mind rather than the opportunity to publish another new contract there is now available a standard form of contract for using with partnering arrangements – PPC 2000 (Figure 5.8). PPC 2000 was developed for individual projects that were to be procured using the partnering ethos. By the nature of the contract it is more suitable to a single project rather than an ongoing framework agreement covering many projects over a period of time, which would need a bespoke contract. PPC 2000 was developed to clarify the contractual relationship where partnerships were being entered into using amended JCT contracts and in a lot of cases this was proving to be confusing resulting in the JCT losing its familiarity. Partnering charters, while proving valuable, were not a legal basis for an agreement and did not outline the system for non-adversarial approach. Drafted by Trowers and Hamlins, a leading London law firm, PCC 2000 was launched in November 2000 by Sir John Egan and published by the ACA. It includes radical ideas set out in the CIC Partnering taskforce guide, launched by Nick Raynsford. There is a standard adaptation for use in Scotland.

The key features of PPC 2000 are designed to encourage a team-based approach between client, professional team, contractor and specialists. It is a single multi-party agreement covering all significant aspects of and the entire duration of

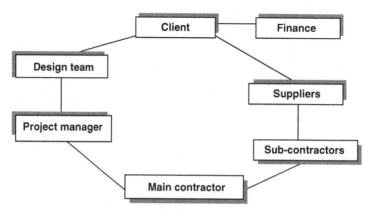

Figure 5.8 Multi-party PPC 2000.

the procurement process. The single contract approach:

- reduces the danger of gaps or inconsistencies between different documentation,
- sets out clearly the relationships between team members,
- helps move parties away from the confrontational mindset of traditional procurement.

The aim of PPC 2000 is that the team is in contract much earlier than is traditional and for negotiations to be carried out post-contract. The contract still allows for a partnering charter and also covers consultants' appointments and agreements – one contract for the whole. It is claimed that PPC 2000 is written in straightforward language with the flexibility to allow the team to evolve as the project progresses, also that its flexibility allows for risk to be allocated as appropriate and moved as the project develops. However, because the PPC 2000 is such a departure from the more traditional contracts, it has proved a bit of a struggle for those who use it for the first time. The contract can be difficult to interpret, as key elements, such as risk sharing and KPIs are left as headings with the detail to be completed later. To guide new users, in June 2003 a Guide to ACA Project Partnering Contracts PPC 2000 and SPC 2000 was launched.

The key features of PPC 2000 include:

- A team approach with duties of fairness, teamwork and shared financial motivation.
- Stated partnering objectives, including innovation, improved efficiency through the use of KPIs and completion of the project within an agreed time and to an agreed quality.
- A price framework which sets out the contractor's profit, central office overheads and site overheads as well as an agreed maximum price.
- A procedure for dispute resolution hierarchy.
- Commitment to the most advantageous approach to the analysis and the management of risk.
- The ability to take out latent defects and/or project insurance.

Some aspects may cause difficulties, for example, Section 3 of the contract gives the client representative the ability to inspect the financial records of any member of the team at any time subject to reasonable notice and access to members' computer networks and data by each member.

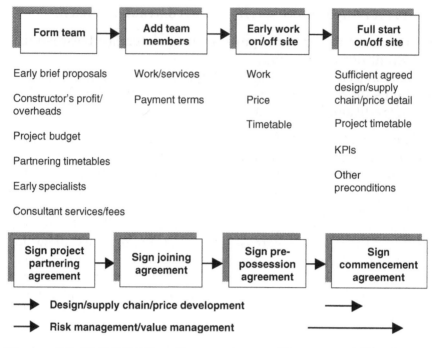

Figure 5.9 PPC 2000 flow diagram (source: Trowers and Hamlins).

The take-up of the PPC 2000 contract is thought to be in the region of between £1 and 2 billion worth of contracts since its launch in 2000. However welcome the introduction of PPC 2000 and partnering, using this approach as a parallel agenda with traditional standard forms of contract as a fall-back position creates confusion and suggests a lack of commitment.

The SPC 2000 for specialist contractors

A further development is the publication of a sub-contract to complement PPC 2000 by the ACA. The specialist contract is intended to provide a standard document so that parties entering into PPC 2000 can have back-to-back arrangements with their sub-contractors or specialists, to use the contract terminology. The SPC 2000 has the same basic structure as PPC 2000, but includes a specialist agreement to which the specialist terms are appended. The specialist contract endeavours to ensure that the constructor and the specialist work more effectively together than is perceived to be the case under the

traditional forms of contract. Unsurprisingly, it prescribes partnering objectives and targets. As with all such provisions, these
are a mixture of aspirations and legally enforceable obligations
and as such there are question marks over the status, enforce-
ability and implications of the aspirational objectives, particularly in the context of the rights and obligations under the
contract. The specialist needs to recognize however that although
it is working in a partnering arrangement with the constructor,
it is also responsible for managing all risks associated with the
specialists works unless otherwise provided for. As part of the
process, the constructor and the specialist must identify what
risks might arise and then share or apportion the risks according to who is more able to manage them.

The NEC Partnering Agreement

To compare the PPC 2000 with the NEC Partnering Agreement
is to compare apples with oranges. Where as PPC 2000 is a free-
standing multi-party contract, governing all the parties' mutual
rights and obligations in respect of a particular project, rather
than being, as in the case of the NEC partnering agreement, an
option bolted to a series of bi-party contracts, which must each
be based on the NEC form. When using NEC option X12 each
member of the partnering team has its own contract with the
client, see Figure 5.10.

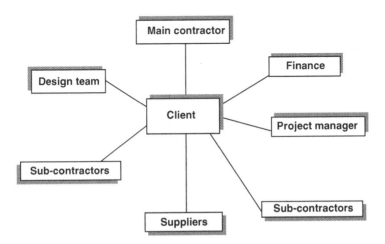

Figure 5.10 Partnering using NEC option X12.

The NEC Partnering Agreement, which by contrast to PPC 2000 is extremely short, acts as a framework for more detailed provisions which must be articulated by the parties themselves in the Schedule of Partners, or in the document called the Partnering Information. It is up to the parties to identify the objectives, further provisions in the Partnering Agreement, set out obligations which are an essential condition if those objectives are to be met. For example, attendance at Partners and Core Group meetings, arrangements for joint design development, risk management and liability assessments, value engineering and value management, etc.

Framework agreements

Framework agreements are increasingly used in both the private and public sectors. Despite many rumours to the effect that European Union (EU) legislation does not permit the use of frameworks in the public sector, in fact they never have been forbidden – provided that it was evident that there was a transparent competition process involving an economic test, then the requirements of the procurement directive were satisfied. Frameworks have been used for some years on supplies contracts, however, in respect of works and services contracts, the key problem has been a lack of understanding as to how to use frameworks, while complying with the legislation, particularly the need for an 'economic test' as part of the process for selection and appointment to the framework.

A framework agreement is a flexible arrangement between parties, which states that works, services or supplies of a specific nature will be undertaken or provided, in accordance with agreed terms and conditions, when selected or 'called-off' for a particular need. *Note*: Inclusion in a framework is simply a promise and not a guarantee of work. Entering into such a framework is not entering into a contract, as contracts will only be offered to the framework contractors, supply chains, consultants or suppliers, as and when a 'call-off' is awarded under the agreement. However the framework does set out the basis of the relationship. In changes being introduced in a widescale overhaul of the EU Public Procurement Directives, programmed to come into force in the UK in early 2005, framework agreements with a duration of 4 years will be able to be used for the procurement of services and works. One high profile initiative using framework agreements is the National Health Service (NHS) ProCure 21.

The new EU Directive outlines the principles that must be satisfied before a framework can operate as follows:

- *Fundamental terms and conditions, including an 'economic test'.* The economic test can take a number of forms, the most common perhaps being a schedule of agreed rates. For example, in the case of NHS ProCure 21, the economic test is satisfied by a process that results in so-called calculation of 'agreed margins', based on the model of business as usual, which are allied to a framework organization's cost models. In addition these agreed margins are also used in ProCure 21 for other matters, such as benchmarking, continuous improvement etc.
- *No renegotiation of terms and conditions.* If the event of a framework contractor being awarded a contract then it should be possible to take a pre-drafted contract without having to renegotiate any of the substantive terms and conditions. It is not permitted under the directive to renegotiate the price basis or any other term the framework members are appointed under. Hence the agreed margins approach adopted by ProCure 21 is considered by NHS Estates far better, as it uses actual cost/open book accounting on specific schemes and can involve the use of standard contracts for schemes etc.

The major advantages of framework agreements are seen to be:

- It forms an inherently flexible procurement tool.
- The avoidance of repetition when procuring similar items.
- Establishment of long-term relationships and partnerships.
- Whenever a specific contract call-off is to be awarded, the public body may simply go to the framework contractor that is offering the best value for money for their particular need.
- Reduction in procurement time/costs for client and industry on specific schemes.
- Enables engagement of supply chain early in procurement process when most value can be added.

National Health Service ProCure 21

NHS ProCure 21 has been constructed around four strands to promote better capital procurement by:

- Establishing a partnering programme for the NHS by developing long-term framework agreements with the private sector that will deliver better value for money and a better service for patients.
- Enabling the NHS to be recognized as a 'Best Client'.
- Promoting high quality design.
- Ensuring that performance is monitored and improved through benchmarking and performance management.

In May 2000 the Department of Health published its report *Sold on Health*, in response to HM Treasury's *Achieving Excellence*. *Sold on Health* had as one of its two terms of reference 'to deliver improvements in the efficiency and effectiveness of the procurement operation and disposal of the NHS Estate'. Objective 5 of the report – *Review of the Capital Procurement Process* introduced the strategy now known as NHS ProCure 21.

The redevelopment of the NHS estate is at the heart of New Labour's political agenda and the NHS Plan and the delivery targets for capital investment. NHS Estates have been tasked with assisting the NHS to procure built assets that will deliver high quality health care to patients including (the following are NHS Plan targets): investing in 100 new hospital schemes by 2010; 20 diagnostic and treatment centres; the refurbishment of 3000 family doctors' premises; 500 (£1 billion) new one-stop primary care centres by 2004; and 25% of hospitals replaced or upgraded by 2013. These are rolled out nationally for all contracts over £1 million in a £3.2 billion per annum building programme.

By any measure a challenge, for in common with most large public sector providers the NHS has suffered from the usual problems of schemes being delivered late, over budget and with varied levels of quality combined with little consideration to whole life costs. Figures from the NHS ProCure 21 *Best Client Manual* (2002) indicated that of 45 NHS major district hospital construction projects completed between 1985 and 1996 that the tender price was exceeded by 10% on 23 of these projects and by more than 20% on 14 of them. In addition 17 of the projects over-ran on time by over 10% and 10 of the projects by more than 20%. Furthermore, one of the main challenges to NHS Estates is the fragmentation of the NHS client base for specific health care schemes, as it comprises several hundreds of health trusts who all have responsibility for the delivery of schemes and each having differing levels of expertise and experience in capital procurement.

Within the North-West and West Midlands areas of the NHS currently covered by a framework there are some 150 separate NHS Trusts and in total there are over 500 NHS entities across the whole NHS, each with the responsibility to procure capital schemes. In addition the NHS faces the problem that, from a procurement perspective, it operates within the heavily prescriptive and regulated public sector.

NHS Estates solution to these problems was to pilot an approach to procurement known as NHS ProCure 21 (the 21 stands for the 21st century) as a radial departure from traditional NHS procurement methods and its cornerstone of the massive capital investment programme in the NHS in the period up to 2010. As with so many new initiatives, the catalyst for this new approach was the Latham/Egan reports of the 1990s. There is nothing particularly radical to students of Latham and Egan about the ethos the NHS intends to implement in its NHS ProCure 21 initiative; a concentration on value rather than cost, integration rather than fragmentation, benchmarking, etc. Framework partnering or ProCure 21 is one of the two procurement paths currently preferred by the NHS, the other one being the previously discussed, The Private Finance Initiative. The principle underpinning the ProCure 21 programme is that of partnering with the private sector construction industry. The key areas earmarked by NHS Estates for improvement by embracing partnering principles are:

- quality,
- value for money,
- time,
- predictability,
- whole life costs.

The NHS ProCure 21 Pilot Framework Agreements for the North-West and West Midlands came into effect in April 2002 and initially will run for 4 years. NHS Estates is currently involved in establishing frameworks for the NHS that will cover all non-PFI Funded projects with a works cost in excess of £1 million. In December 2002 the procurement process to establish frameworks to cover other areas of the NHS was commenced. The procurement process for these frameworks complies with EU legislation, i.e. they were advertised through-out the EU via the Official Journal. Before the introduction of NHS ProCure 21, in order to comply with EU legislation, the NHS was obliged to:

- undertake a competitive tendering process for schemes not in excess of the EU threshold value or
- to issue notices in the *Official Journal of the European Commission* (OJEC) for all schemes that exceeded the threshold value for the application of the frameworks.

Following the establishment of the frameworks NHS Trusts do not have to go through a tendering or OJEC procurement process, saving considerable time and money and enabling them to progress schemes more quickly and engage the supply chains to work with them at the earliest possible stages. *Note*: NHS Estates are focusing on all health care facilities from Acute to Primary Care so it is the procurement/delivery process that will be similar, not necessarily the building type. Frameworks can be used in repetitive contracts and eliminate the need to call for tenders for similar types (Figure 5.11) of services or materials under the EU Directive's multi-supplier frameworks follow two procedures:

- In the first, the supplier for each order is selected solely on the basis of the original bid;
- In the second, suppliers have the opportunity to amend and complete bids before the winner is selected for each order;
  - ProCure 21 uses the Type 2 Framework procedure, that is to say Principal Supply Chain Partners (PSCP) sit on a pre-tendered framework and are drawn off by partici-pating NHS Trusts without the need to OJEC tender.
  - Once included within a framework, the choice of PSCP is made by individual NHS Trusts on the basis of 'can we work with this team?'

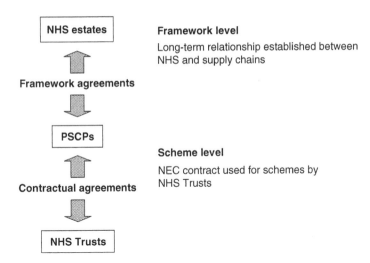

Figure 5.11 ProCure 21 framework arrangements.

The key differences between the PFI and ProCure 21 are:

- PFI is project based, whereas ProCure 21 is more flexible although the batching of smaller PFI projects. The first batched hospital deal being announced in May 2003, goes some way to addressing this difference.
- ProCure 21 Partnering is based over a range of projects, where as the PFI approach is specific to one project.
- Continuous improvement is embedded into ProCure 21 partnering and not necessarily into PFI project.

The basic principles

Earlier the principles of framework agreements were described, the way in which those principles are interpreted by NHS ProCure 21 are as follows.

In the roll-out of ProCure 21 the framework consists of five PSCPs. A PSCP is defined as an organization that:

- can take on single point responsibility to manage the design and construction of facilities;
- has specifically assembled a supply chain with experience of working in long-term partnering arrangements with key suppliers;

Figure 5.12 Best client competencies as defined by ProCure 21.

- can deliver publicly funded and/or privately financed schemes included within the scope of the NHS Procure 21 Scheme range.

Throughout the roll-out of ProCure 21 heavy emphasis was placed on turning the NHS into a best practice client, that is to say, not leaving it to the supply side to deliver all the improvements. This is being achieved through a series of training programmes to produce Project Directors – the overall client advisor. Figure 5.12 indicates the competencies identified by the NHS Estates for the development of best clients.

Best client

Considerable emphasis in the ProCure 21 process has been placed on best practice client collapsed to 'best client' for ProCure 21. NHS Estates fully recognize that for change to take place that both demand and supply have to contribute and so in June 2002 the first 279-page *Best Client Manual* was issued. Best client approach is based on a win/win approach which ensures that the benefits to the NHS are not won at the expense on the supplier base. What are considered to be

Best Client attributes? According to the *Best Client Manual* these need to be largely determined from the supplier's point of view but include the following:

- Reduces red tape and bureaucracy by having clear expectations and the ability to measure what is important and resolve disputes.
- Takes a long-term view to mutual development and learning and is welcoming to new ideas.
- Has a focus on quality and value for money, operating to the highest quality standards taking whole life costs into account.
- Shows trust and willingness to partner and share sensitive information.
- Is committed to open communication and operates to compatible values.
- Is committed to increasing the speed of response through the use of technology and shared processes.
- Shares rewards equitably by paying on time and seeks the potential for higher margins.
- To achieve these goals Accredited Project Directors will be drawn from the existing NHS management and trained as accredited client advisors to assist Trusts in becoming 'Best Clients.'

Principal Supply Chain Partners

Similar to a prime contractor the term was deliberately chosen to try to reinforce the message that PSCPs do not need to be contractors – they just need to be organizations that take responsibility for delivery by supply chains – Defence Estates are trying to get this message across as well. NHS Estates felt that to avoid the use of the term contractor, that has certain connotations within the construction industry, might help. A PSCP therefore is an organization with established supply chains capable of providing all necessary resources to design, build, operate and in some cases finance, NHS capital schemes. In addition they should have a proven ability successfully to engage in partnering agreements and effectively manage the supply chain. Currently five PSCPs have been appointed for the North-West and West Midlands Area but for the National Frameworks there will be 8–12 PSCPs with supply chains. Trusts can select one from the list or more if they want to but no price is involved; it is merely an empathy process in terms of the team presented for the job and what added value they can bring to the Trust based on preliminary project proposals. Once selected, the Trusts work with the PSCP during the development of the outline and full business cases to arrive at the guaranteed maximum price (GMP). If for some reason the Trust and the PSCP cannot agree on a GMP then the Trust is free to approach another PSCP from within the framework, with the original PSCP being financially compensated for the work already done – but for this it must be emphasized, is a last resort.

Initially a standard form of scheme agreement will be used such as PPC 2000 or the NEC Option X12 for publicly funded schemes. A PSCP's primary supply chain will consist of seven Principal Supply Chain Members (PSCMs) for publicly funded schemes and eight PSCMs for PFI schemes and must include:

- Architectural design
- M&E services design
- Cost management
- Constructor (large schemes)
- Constructor (small schemes)
- M&E Services installer
- Facilities management (hard FM)
- Facilities management (soft FM–PFI Schemes only).

The remaining PSCMs (9–11) are at the discretion of the PSCP (see Figure 5.13).

Both PSCPs and their PSCMs will be asked to provide evidence of, and commitment to, best practice in partnering and supply chain management. Initially it was proposed to establish specialist procurement teams made up from external consultants, with the role of advising Trusts on procurement matters. However that function has now been subsumed within the PSCP supply chain. The demands placed on prospective prime contractors by NHS Estates to the criteria for inclusion on a short list for traditional single stage competitive tendering makes an interesting comparison.

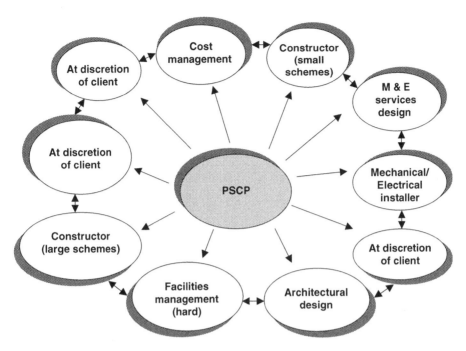

Figure 5.13 PSCP – ProCure Model.

The procurement process

The selection procedure for appointing PSCPs to a ProCure 21 Framework Agreement is summarized below. The procurement process involves two stages. All candidates are responsible for all costs incurred in connection with their responses to all stages of the appointment process.

Stage 1 – Pre-qualification

A complete pre-qualification form is completed and returned to NHS Estates for evaluation. NHS Estates manage the selection process to identify twelve candidates to proceed to the Final Selection Process (FSP).

The pre-qualification process has two parts. The first part asks for financial and economic information about the PSCP, their PSCMs and their constituent parts. The second part looks in detail at the technical and organizational capacity of the PSCPs and their PSCMs.

Stage 2 – Final Selection Process

Twelve candidates will be invited to participate in the FSP for each framework agreement. The basis of the selection will be an assessment by the NHS Estates evaluation team on the basis of the candidates' responses to:

(a) the completion of an economic test,
(b) site and/or office visits,
(c) interview.

Cost models (the economic test)

The economic test for framework selection will cover all whole life cost aspects of capital procurement through cost model documents specifically created to suit the relevant framework, and will cover the following points:

- Design
- Construction
- Hard FM
- Soft FM (PFI only)
- Risk
- Finance (PFI only).

The cost models are based upon real projects and have each aspect costed in order to act as a benchmark against which the PSCPs will bid. The costs submitted by the PSCPs at the framework stage will act as benchmarks to test the appropriateness of the costs arrived at for the scheme tender stage of the relevant scheme.

These benchmarks are key to the assessment of progress in reducing costs against targets. PSCPs will be required not only to give their assessment of costs against the models for the first year, but also to indicate how they expect the benefits of partnering and supply chain management to deliver annual improvements over the life of the framework. These benchmarks will therefore allow the progress of the PSCPs to be judged against a pre-agreed programme for improvement; and will form the backbone of benchmarking for use with future frameworks and schemes.

Site and office visits

This will provide an opportunity for NHS evaluators to establish how information provided at PQQ is being implemented in a 'live' environment. This will cover the whole supply chain, clients, policies, procedures, training, human resources, health and safety, etc.

Details of the key personnel who will be involved in negotiating scheme agreements will be required if these are different from those responsible for the application to the framework agreement.

Full details of all projects undertaken in the last 3 years, whether proceeding or cancelled will be required together with contact details.

The selection process

Evaluation guidelines are issued to the candidates at that time and selection is based on the ability to demonstrate:

- Technical:
  - Proven track records of supply chain members' partnering experience with clients and supply chain management expertise.
  - Quality of resources and expertise available.
  - Design capability.
  - Management contracting capability.
  - Hospital commissioning ability.
  - Facilities Management track record.
- Commercial:
  - Ability to manage time, cost, quality and risk.
  - Ability to put appropriate finance in place.
  - Economic test and year-on-year improvement.
- Soft:
  - Evidence of ability to partner.
  - Evidence of supply chain management ability.
  - Evidence of appropriate attitude and culture.
  - Understands NHS culture.
  - Proven ability to innovate
  - Willingness to take part in continuous improvement.

Response to the pilot projects was encouraging. A call for interested parties to become one of the five PSCPs was placed in the

OJEC. As a result 222 organizations replied and these were initially reduced to twelve and then eventually to five. The five PSCPs entered a 4-year framework agreement.

Finally, a particular issue in the NHS is standardization. By their own admission NHS Trusts seem unable to standardize at various levels, ranging from procedures to components and this can result in wasted time, effort and subsequently money. It may and also frequently does mean that a standard six-bed ward in one hospital is vastly different to a six-bed ward in another, regardless of them providing identical facilities. One way of trying to overcome this problem currently being investigated is the construction of demonstration wards as a best practice solution.

Case study – Milton Keynes General Hospital (MKGH)

To date there are no completed ProCure 21 health projects; however Milton Keynes General Hospital, a diagnostic and treatment hospital, gives some clear indications of the potential benefits of this new approach to procurement.

The project is a £12 million 60-bed hospital with four new operating theatres; the scheme was announced by the government in December 2002 with a start on site in May 2003 and completion by the end of 2004. The tight programme is deliverable only because of the framework agreements already in place and the pre-qualified PSCP was able to commence work quickly. In this case the Trust were able to appoint a PSCP in 4 weeks by using an open day and interview process for all PSCPs in the framework agreement, under conventional procurement processes this would never have been possible. In addition to the saving in appointing a contractor, spin-offs from the ProCure 21 approach have been judged to be:

- High quality design due to the detailed knowledge and expertise of the PSCP.
- Technical knowledge and competence developed within MKGHT together with confidence developed with the clinicians.
- The PSCP retained Trust consultants, to ensure these benefits were not lost.
- PSCP had the flexibility to employ members of existing teams alongside their own people.

- Provision of information from other P21 projects to trust – networking.
- Efficient reuse of existing information, saved time and money and avoided reinvention of the wheel.
- PSCP's knowledge of hospital used to add value to non-project specific uses.
- Resultant enhanced knowledge will improve the project.
- PC21 scheme provides the framework and back-up to ensure the working relationship is auditable as well as functional.
- It is expected that the current spirit of flexibility and openness will continue throughout the programme, ensuring that the hospital is an excellent project for the long term.

ProCure 21 is still in its infancy; however early indications are that the approach adopted by NHS Estates is going to prove successful.

Tradition with a twist

Traditional lump sum procurement, based on bills of quantities, still accounts for approximately 40% of all UK construction activity. The following section is a review of some traditional procurement strategies including design and build (D&B):

- two-stage competitive tendering,
- a new approach to single stage competitive tendering,
- D&B and variants.

Two-stage competitive tendering

First used widely in the 1970s, as the title suggests this procurement path is based on the traditional single stage competitive tendering approach, that is bills of quantities and drawings being used to obtain a lump sum bid. However, there are a number of significant differences that have led to it being labelled 'fast track'. The two-stage approach breaks the traditional mould in some respects in that, for it to be truly successful, the system relies on the establishment of trust between the client and the contractor and for this reason this approach has been dismissed by some in the past as a difficult option. In fact anecdotal evidence on the efficiency of two-stage tendering varies widely. Nevertheless, the advantages include early contractor involvement, a fusion of the design/procurement/construction phases and a degree of parallel working that reduces the total procurement and delivery time. A further advantage for the unadventurous is the fact that the documentation is based upon bills of quantities and therefore should be familiar to all concerned. A disadvantage identified by *Masterman* is, when using this approach that early price certainty is ruled out, as the client can be vulnerable to any changes in level in the contractor's pricing, between the first and second stages. Although this practice cannot be ruled out it surely reinforces the point made at the start of this paragraph when, the need for trust between parties was emphasized as an essential ingredient for success and not just for two-stage tendering but procurement *per se*.

The approach when using this procurement strategy is as follows:

Stage 1

The decision having been taken to proceed with the project the quantity surveyor begins to prepare the basis for the first-stage bid. The first-stage tender is usually based on approximate bills of quantities, however this does not have to be the case, other forms of first stage evaluation may be used, for example a schedule of rates, although there is perhaps a greater degree of risk associated with this approach. As drawn information, both architectural and structural, is very limited, the choice of bid documentation will be influenced by the perceived complexity and predictability of the proposed project. The two-stage approach places pressure on designers to take decisions concerning major elements of the project at an earlier stage than normally. A common feature with many non-traditional procurement systems including, Design and Construct and PPP, is that the design development period is truncated and that as a result, the design that is eventually produced can lack architectural merit. Assuming that bills of approximate quantities are being used they should contain quantities that reflect:

- items measured from outline drawings,
- items that reflect the trades, it is perceived, will form part of the developed design,

- items that could be utilized during the pricing at second stage.

It goes without saying that the preparation of the first stage documentation is a situation where the traditional quantity surveying skills of, measurement and expertise in construction technology come into their own. At an agreed point during the preparation of the first stage documentation the design for the first stage is frozen, thereby enabling the approximate bills of quantities to be prepared. Without this cut-off point the stage 1 tender documentation could not be prepared. Note, that it is important to keep a register of which drawing revisions have been used to prepare the stage 1 documentation for later reference during the preparation of the firm stage 2 bills of quantities. However, while first-stage documentation is prepared the second-stage design development can continue. During the first-stage tender period attention should be focused on the substructure, as it is advantageous if at the first tender stage this element is firm. If a contractor is selected as a result of the first-stage tender then they may well be able to start on site to work on the substructure while the remainder of the project is detailed and the second-stage bills of quantities are prepared and priced.

Once completed the first-stage approximate bills of quantities together with other documentation are despatched to selected contractors with instructions for completion and return in accordance with normal competitive tender practice. On their return, one contractor is selected to proceed to the next phase. It should be noted that selection at this stage does not automatically guarantee the successful first-stage bidder award of the project, this is dependent on the stage 2 bidding process.

At this point the trust stakes are raised – assuming that a contractor is selected as a result of the stage 1 tender the following scenarios could apply:

Client	*Contractor*
The client trusts the contractor to be fair and honest during the stage 2 negotiations. Failure could result in the client having to go back to the start of the process.	Although selected by stage 1 process – no guarantee of work. No knowledge as to accuracy of first-stage documentation or client's commitment to continue.

Client could ask contractor to join design team to assure buildability.	Contractor could be asked to start on site on substructure while stage 2 is progressed.
Client relies on the design team to prepare documentation for stage 2 timeously. If stage 2 bills of quantities are not accurate the work will have to be remeasured for a third time!	Contractor rewarded on the basis of letters of intent, quantum merit, etc.
The client could, under a separate contract, engage a contractor to carry out site clearance works while stage 1 bids are evaluated.	Contractor could exploit position during stage-two pricing.

Stage 2

The purpose of stage 2 is simply to convert the outline information produced during stage 1 into the basis of a firm contract between the client and the contractor, as soon as possible. With a contractor selected as a result of the first-stage process pressure is placed on the design team to progress and finalize the design. Between contractor selection and stage 2, usually a matter of weeks, the design team should prepare and price the second-stage bills of quantities. During this phase it is usually the quantity surveyor who is in the driving seat and he/she should issue information production schedules to the rest of the design team. As design work on elements is completed it is passed to the quantity surveyor to prepare firm bills of quantities which are used to negotiate the second-stage price with the contractor on a trade by trade or elemental basis. It is therefore quite possible that the contractor will be established on site before the stage 2 price is fully agreed. Unless the parameters of the project have altered greatly there should be no significant difference between the stages 1 and 2 prices. Once a price is agreed a contract can be signed and the project reverts to the normal single-stage lump sum contract based on firm bills of quantities; however the adoption of parallel working during the procurement phase ensures that work can start on site much earlier than the traditional approach. Also the early inclusion of the main contractor in the design team ensures baked-in buildability and rapid progress on site.

Critical success factors

- Trust between the parties.
- The appointment of a contractor that can innovate and is prepared to contribute to the design development.
- Information production keeping up with requirements, failure to achieve this can lead to a good deal of embarrassment.

The UNCITRAL Model Law referred to previously in Chapter 4 suggests the following approach to two-stage tendering:

- The first stage is used to ask suppliers or contractors to submit their technical proposals but without a tender price.
- The client enters into discussion with the supplier/contractor and those who appear to fall short of the criteria are discarded. At the same time the client draws upon the proposals of the first-stage tenders to prepare the basis for the stage 2 documents.

The remaining contractors are then asked to submit tenders based on the revised documentation.

Single-stage selective tendering

A twist to the traditional single-stage selective tendering themes:

- A fully developed business case is subjected to rigorous cost planning by consultants in conjunction with the client.
- Suitable contractors, say 2 or 3, are pre-selected using appropriate selection criteria, such as:
  - Commitment to supply chain management
  - Ability to guarantee life cycle costs
  - Capability to deliver
  - Etc.
- The pre-selected contractors, are asked fully to cost project proposals and submit their Best and Final Offer (BFO).
- On the basis of the BFO a contractor is selected and enters into contract guaranteeing the BFO price and in some cases whole life costs for a pre-determined period.
- Work starts on site.

Ironically, each of the contractors in this process will probably have to use the services of an independent quantity surveyor to prepare bills of quantities, the reason that, many years past, was the principal reason for the rise of the quantity surveyor.

D&B variants

D&B or Design and Construct is a generic term for the following approaches to procurement:

- *Traditional D&B*: The contractor is responsible for the complete design and construction of the project.
- *Enhanced D&B*: The contractor is responsible for the design development, working details, as well as construction of the project.
- *Novated D&B*: The contractor is responsible for the design development, working details and supervising the subcontractors, with assignment/novation of the design consultants from the client.
- *Package Deal and Turnkey*: The contractor provides standard buildings or system buildings that are in some cases adapted to suit the client's space and functional requirements.

The procurement path discussed in this section will be limited to traditional D&B that together with PPP/PFI is one of the procurement systems currently favoured by many public sector agencies. During the past decade the use of D&B variants in all sectors has increased from 11% in 1990 to 40% in 2000 (Davis Langdon and Everest/RICS, 2000), the reasons are clear. D&B gives a client the opportunity to integrate, from the outset, the design and the construction of the project. The client enters into a single contract with one company, usually a contractor who has the opportunity to design and plan the project in such a way as to ensure that buildablility is baked into the design. With specialist involvement from the start, this approach promises a shorter overall delivery time and better cost certainty than traditional approaches. The performance of D&B has been the subject of studies by Konchar and Sandivo (1998) and Bennett *et al.* (1996). The results of the studies indicated that D&B outperforms traditional forms of procurement in several respects; however the differences are not that significant.

One reason for this could be that, within the UK, for the organizations that provide D&B services, this is not their key competence and therefore when the opportunity comes to bid for a D&B project, temporary organizations of designers and constructors have to be formed specifically for a project. For the contractor and the designers the next project may be traditional contracting and therefore the temporary organization is disbanded. Both studies mentioned above concluded that the total delivery speed of D&B compared with traditional approaches is 30–33% faster. Also the percentage of projects that exceeded the original estimate by more than 5% was 21% in D&B compared to 32% for traditional procurement. It would seem then that although delivery times are shorter when using D&B that improvements in cost certainty are only marginal.

The main criticisms of D&B procurement are centred around the lack of control over quality of design, with little time being allocated for design development and possible compromises over quality to provide cost savings by the contractor.

Successful use of D&B relies on the contractor preparing proposals that include:

- a contract sum analysis that itemizes the financial detail on an elemental basis and
- detailed proposals of how the requirements of the client's brief will be satisfied.

D&B is recommended by the Office of Government Commerce for procurement within a partnership arrangement.

Postscript: recent changes in EU procurement directives

European public procurement is one of the largest procurement markets in the world and just like the rest of the European dream, European public procurement is taking big steps, a little at a time. The Lisbon European Council adopted an economic reform programme with the aim of making the EU the most competitive and dynamic knowledge based economy in the world by 2010. A key part of this programme is to ensure that the internal market works efficiently for the procurement of works and services. 2005 will see the biggest shake-up in the EU public procurement directives since their original drafting in

the 1980s. The flaws in the procurement system for construction
and engineering works is well documented, such as the inability
of the EU effectively to punish the member states who break
the rules. A recent report by Directorate General XV of the
European Commission concludes that it is not only on works
contracts that EU states are still operating a buy national policy, services, that professional services, such as project management, etc. are subject to the same discrimination when it comes
to selection on cross border procurement opportunities.

The report published in July 2002 covered a large variety of
activities such as consultancy services, estate agents, engineering, construction, etc. emphasizes that the consumers are the
principal victims of the dysfunctioning of the internal market.
Further the report concluded that services are much more prone
to internal market barriers than goods and are harder hit and
that this occurs at every stage of the business process from establishment in another member state to promotion and marketing.
The nature of the barriers are broadly catagorized into the more
obvious ones such as legal and technical as well as new barriers
such as discretionary powers and non-transparent procedures
operated by some member states. The report, which is to be followed by a second study to be published in 2003, see the dissolution of barriers to services as a major priority for the EU.

Tender evaluation performance will be discussed in the following chapter.

Bibliography

Audit Commission (2003). *PFI in Schools, the Quality and Cost of
Building and Services Provided by Early Private Finance Initiative
Schemes*. HMSO.

Barlow J., Cohen M., Jashapara A. and Simpson Y. (1997). *Towards
Positive Partnering*. Policy Press, Bristol.

Bulter E. and Stewart A. (1996). *Seize the Initiative*. Adam Smith
Institute.

Crates E. (2002). NHS Procure 21: a new route to better building.
RICS Business, November/December 30–32.

Grahame A. (December 2001). *House of Commons Research Paper
01/17, The Private Finance Initiative*. House of Commons Library.

HM Treasury (April 2003). *The Green Book*. HMSO.

Hemsley A. (2003). A guide for the perplexed. *Building* 20th
June, p 53.

Howell G. *et al*. (2002). *Beyond Partnering: Towards a New Approach to Project Management.*

House of Commons Committee of Public Accounts (2003). *Private Finance Initiative: Redevelopment of MOD Building.* The Stationery Office.

Masons (2002). *Partnering – The Industry Speaks.* Masons.

National Research Council (2000). *Surviving Supply Chain Integration.* National Academy of Sciences.

Pollock A. *et al*. (2002) *Private Finance and Value for Money in NHS hospitals: a policy in search of a rationale?,* British Medical Journal, 18 May, Issue 324, pp 1205–1209.

Ross J. (1999). *Project Alliancing in Australia – Background, Principles and Practice.* (Presentation to an industry summit on relationship contracting in Sydney.)

United Nations Assembly (1993). *UNICTRAL Model Law*, UN, New York.

Webb A. (1999). Alliancing for capital projects: one way for the future. *The Engineering Management Journal* 9 (4 and 5).

Web sites

www.audit-commission.gov.uk
www.parliament.uk
www.bmj.com
www.nhsestates.gov.uk
www.procure21.gov.uk

6

Procurement – performance and tender evaluation

Generally

Throughout this book the traditional practice of assessing tenders on the basis of lowest price has been called into question, as this approach has been found to ignore the selection of best value solutions, as well as largely ignoring whole life cost implications. The widespread adoption of procurement systems that focus on value, rather than cost however bring with them their own sets of problems. For example, how does one objectively determine whether a bid A delivers better value than bid B, when the criteria for measuring value can be so diverse? Assuming that a team has been tasked with the procurement of a new built asset using best value principles, how can it be proved at the end of the day that best value is just that? How is innovation assessed? A further complication is the use of non-prescriptive documentation, placing the responsibility on the tendering organizations to design and put forward the best VFM solutions. Perhaps more than most countries, the UK in particular, which has for decades been wedded to prescriptive approaches and standard documentation, has a steep hill to climb, when it comes to assessing bids not just on the basis of price alone. First, it is necessary to define what best value represents in a particular project and then measure it. In the 'good old days' when most contracts were let on lump sum competitive tendering, the client could sit back at the end of the day knowing that the winning bid was, if nothing else, the cheapest price that had been submitted on the day.

Concepts of value

As discussed in Chapter 1, value has many definitions, one of which is that value relies on the relationship between the satisfaction of many differing needs and the resources used in doing so. Therefore, the fewer the resources used, or the greater the satisfaction of needs, the greater the value.

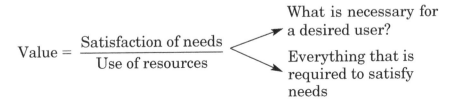

$$\text{Value} = \frac{\text{Satisfaction of needs}}{\text{Use of resources}}$$

What is necessary for a desired user?

Everything that is required to satisfy needs

However a further complication is that stakeholders, internal and external customers, etc. may all hold differing views of what represents value. For example, some of the stakeholders in a typical medium sized project are listed in Table 6.1, each perhaps with a different perception of value.

Of course, relative value may be improved by increasing the satisfaction of need, even if the resources used in doing so increase, provided that the satisfaction of need increases more than the increase in use of resources. In an attempt to define a little more simplistically, value in the construction context, the Office of Government Commerce defines the government's VFM procurement policy, as 'best value for money is the optimum combination of whole life cost and quality to meet the customer's requirements.'

Concepts of performance

Some modern approaches to procurement are turning the traditional strategies on their head. For example, as has already been stated, if the relationship of capex to whole life costs is 20:80, is it worth evaluating a bid on anything other than a whole life costs basis? Instead of selection being based on bills of quantities and bottom line figures at the end of the design stage, the selection of a contractor is increasingly being made before the project details are even designed. In the previous chapter, the use of frameworks was discussed, and

Table 6.1 Project stakeholders

Design	Construction	Performance in use
Client	Main contractor	Client
Project manager	Domestic	In-house management
Safety/quality	subcontractors	Maintenance
consultant	Nominated	contractors
Architects	subcontractors	Facilities consultant
Civil designers	Project manager	Project manager
Mechanical	Material suppliers	Safety/quality
consultants	Plant/equipment	consultant
Electrical	suppliers	Civil engineers
consultants	Architects	Structural engineers
Specialist designers	Financial institution	Mechanical engineers
Quantity surveyor	Insurance agency	Electrical engineers
Financial institution	Regulatory bodies	Quantity surveyor
Insurance agency	Client	Regulatory bodies
Regulatory bodies		Main contractor
		Domestic
		subcontractors
		Nominated
		subcontractors
		Material suppliers
		Plant/equipment
		suppliers

in particular the way in which this strategy is used by the National Health Service (NHS) ProCure 21 process. The pre-qualification for ProCure 21 is so rigorous, that inclusion in a framework could be considered as an evaluation process in its own right. As discussed, the actual selection of a supply chain team for a particular project is carried without negotiation of project-specific prices. Therefore, using this approach the client has the opportunity to set the pre-qualification criteria at a level that reflects the required performance levels over a range of items. In the private sector one of the champions of framework procurement is BAA. In May 2003 BAA announced that it was to embark a second generation of framework agreements with both contractors and consultants. The main differences from the first generation will be less bureaucracy, that is to say a scaling down of the number of questions in the pre-qualification process. Of course, BAA operates within the private sector where the rules, particularly the prescriptive EU Public Procurement Directives do not apply. The specific pre-qualification approach adopted by NHS ProCure 21 will be described later in the chapter; however first consider the general principles of the process of pre-selection of pre-qualification of contractors, which can equally well be applied to selection of consultants.

Pre-qualification of contractors may be used in the procurement process in a number of ways. Either to select suitable contractors and their supply chains to sit on, for want of a better term a framework agreement, to be 'called off' as and when needed. Once selected, and induced in the framework the only matters to be agreed between the client team and the contractor are:

- a guaranteed maximum price,
- a method statement,
- an agreed completion date.

Alternatively, tenders for individual projects may be evaluated on a set of unique pre-selected criteria. The disadvantage of this approach being that criteria have to be set and assessed at project level for each new job, where as with frameworks, the legwork only needs to be done once.

Appointment of a contractor using pre-qualification criteria

The choice of a contractor to carry out a project is divided into two parts:

- the selection of contractor(s), based upon their status and previous performance,
- the award process that examines the proposals for the specific contract.

The eventual selection and appointment procedures will of course become much easier and much more reliable if organizations have been pre-selected against a set of objective criteria before being included on any sort of list, either long or short.

1. *The selection process*

The selection of a short list of contractors to carry out a specific project can of course be in accordance with any criteria deemed to be important to the project team and the client. However, there should be an attempt to base selection on objective criteria rather than on the basis that the contractor is available. The selection process could follow these steps:

- Establishment of selection criteria – see later section.
- The weighting of the selection criteria.
- Where appropriate – the thresholds for selection.
- A selection mechanism.

Selection criteria

Many quantity surveyors will be familiar with the process of drawing up a tender list for a contract to be based on a lump sum bill of quantities. Established practice was that to ensure competitive bids, it was imperative that at least six contractors should be included, all of whom should meet with the approval of both the architect and the client. A difficult task at the best of times, but during a period of full order books, downright difficult. It often was and still is the case that contractors were included merely because they were available and of course the question had to be asked – why was this when everyone else was busy? The criteria for pre-selection of a contractor for a tender list vary according to the nature of the project, but generally they should be aimed at eventually selecting the most suitable organization capable of carrying out the required work while taking into account the need to achieve value for money.

There follows a few examples of criteria that could form the basis of pre-selection:

1. Status of the bidders including:
    - financial standing,
    - credit rating,
    - insurance provisions.

2. Team working arrangements in place within the organizations including:
   - supply chain management expertise,
   - partnering experience.
3. Ability to manage the type of project being bid for including:
   - qualifications and experience of team members.
4. Technical capability including:
   - experience of similar projects.
5. Functional aspects of the type of project being bid for including:
   - appreciation of whole life costs and sustainability issues,
   - demonstration of the ability to deliver innovation in projects.

In order to make the selection more objective each of the selection criteria can be weighted to indicate the relative importance of each section. During the selection process each candidate organization is assessed and then awarded a score against each criteria, for example see Table 6.2.

Teamwork arrangements

At the end of the process the weighted scores are totalled, considered and the most suitable organizations are included on the tender list.

In addition, for each criterion, client and project teams need to determine the minimum requirements for acceptability. Failure to meet such standards should preclude an organization being considered further. Minimum standards for participation should match the corporate philosophy of the client body, for example the consideration of ethical or sustainable issues, but should not be set at unreasonable levels as this may unfairly disadvantage small and medium sized enterprises.

Table 6.2 Selection criteria

Selection criteria	Criteria weighting	Score	Weighted score
Ability to demonstrate highly integrated and well managed supply chains	30%	60	18

There also needs to be a quality threshold under the heading of technical capability, once again set at a level that corresponds to the client/corporate body strategy. Having drawn up and weighted the criteria the selection process can begin. It is important of course to inform potential tender list candidates of the range of information and skills necessary to be demonstrated during the selection process. Having decided which organizations are to be included on the tender list the next stage is the preparation of the tender documentation. In addition to general information relating to the preparation and return of tenders, it is essential that the award criteria should be fully defined in the bid documents. Once more these must be appropriate, specific to the particular project and relevant to assessing whether the bids provide value for money. The criteria should be set up before asking for expressions of interests. The items included in Table 6.3 and in the following section, the award process, provides some examples of the criteria for selection.

2. *The award process*

The award process begins with the examination of the bids from the organization in the tender list. Every tenderer should

Table 6.3 Quality scores

Quality criteria	Criteria weighting	Contractor A		Contractor B	
		Score	Weighted score	Score	Weighted score
Innovation	9	50	4.5	40	3.6
Partnering	7	40	2.8	60	4.2
Risk management	11	45	5.0	30	3.3
Project organization	5	35	1.8	70	3.5
Aesthetics	5	50	2.5	85	4.3
Programme	12	45	5.4	50	6.0
Functionality	22	65	14.3	70	15.4
Qualifications	5	60	3.0	55	2.8
CDM	9	80	7.2	50	4.5
WLC	15	60	9.0	70	10.5
Quality score total	**100**		**55.5**		**58.1**

Project quality weighting: 60, Project price weighting: 40.

be informed of the basis by which bids will be judged and awarded. When the tenders are received organizations may be assessed on how well they satisfy the award criteria. The relative importance of criteria should be weighted, as for example shown in Table 6.3.

Quality/price ratio

The first task is to establish the quality/price ratio for the project. This is an attempt to evaluate the bids while taking into account the optimum combination of costs and quality, which will vary according to the nature of the project, with quality generally being given a higher rating the more complex and innovative the project.

Award criteria

Table 6.3 is an example of a structured approach to tender evaluation which takes into account some of the aspects covered in the preceding paragraphs and weights the various aspects, thought to be important criteria of the new built asset. It is an attempt to come to an objective decision on the merits of price and quality. For simplicity only two contractors have been included in this example which has been based upon the *Client Pack for Construction Works Procurement*:

- Contractor A – tender price: £1,835,000.00
- Contractor B – tender price: £2,134,532.00

Before the tendering process commenced, the quality criteria, identified by the project team, possibly as a outcome of a value management workshop, were made known to the contractors, together with the relative weighting of each items as shown below. The contractors are assessed on how well they satisfy the award criteria, including any mandatory components. As part of the evaluation process each of the bids was scored by the project team. It is assumed that both bidders reached the appropriate quality threshold for each of the criteria. For this project the weighting of Quality to Price was set as 60:40. Each item is scored on a scale of 0–100, with 100 being the top score.

Evaluation of price

There are a number of options for evaluating the price. The system used in example Table 6.3 is as follows:

- The mean price of the tenders is allocated the score of 50.
- One point is deducted from the score of tenderers for each percentage point above the mean and vice versa.
- Project quality weighting: 60.
- Project price weighting: 40.

Price scores

The price score for each contractor is measured against the average of tender prices, that is in this case – £1,984,766.00:

- *Contractor A*
  - Therefore the price score, adjusted against the mean, see above for calculation, is 58
  - Price weighting × tender score
  - $58 \times 40\% = 23.2$
- *Contractor B*
  - Price score, ditto = 42
  - Price weighting × tender score
  - $42 \times 40\% = 16.8$

Quality scores

Quality weighting × Quality score, from Table 6.3:

- Contractor A: $60\% \times 55.5 = 33.3$
- Contractor B: $60\% \times 58.1 = 34.9$

The evaluation process is now completed by adding together the price and the quality scores as follows:

- Contractor A: $23.2 + 33.3 = 56.5$
- Contractor B: $16.8 + 34.9 = 51.7$

The bid with the higher score, Contractor A, is closest to the pre-selected criteria and therefore is judged the more competent to carry out the work.

Tender evaluation

A variation on the traditional approach to tender evaluation is a two envelope tendering system. This approach has the advantage of allowing the quality element of the bid to be evaluated without influence by the price. In a system of this type, tenderers are instructed to submit their tenders in two envelopes: Envelope A contains the organization's quality element of the tender and Envelope B, the price element. The tenders are then considered in the order A then B. When using this method great care should be taken to ensure that, for all bids that comply with the minimum quality requirement, Envelope B is opened.

Selection criteria

How are the levels determined and in some cases measured for the selection process; it may be by reference to an organization's own criteria or in some case to industry or competitor's standards. Two approaches to setting standards are now discussed and as it will be seen they are not without their problems.

Benchmarking

Benchmarking is a generic management technique that is used to compare performance between varieties of strategically important performance criteria. These criteria can exist between different organizations or within a single organization provided that the task being compared is a similar process. It is an external focus on internal activities functions or operations aimed at achieving continuous improvement (Leibfried and McNair, 1994). Construction, because of the diversity of its products and processes, is one of the last industries to embrace objective performance measurements. There is a consensus among industry experts that one of the principal barriers to promoting improvement in construction projects is the lack of appropriate performance measurement and this was referred to also in Chapter 2, in relation to whole life costs calculations. For continuous improvement to occur it is necessary to; have performance measures which check and monitor performance, to verify changes and the effect of improvement actions, to understand the variability of the process and in general it is necessary to have objective information available in order to make effective decisions. Despite the late entry of benchmarking to construction this does not diminish the potential benefits that could be derived, however it gives some indication of the fact that there is still considerable work to be undertaken both to define the areas where benchmarking might be valuable and the methods of measurement. The current benchmarking and key performance indicator (KPI) programme in the UK construction industry has been headlined as a way to improve underperformance. However despite the production of several sets of KPIs large-scale improvement still remains as elusive as ever. Why is this?

The Xerox Corporation in America is considered to be the pioneer of benchmarking. In the late 1970s Xerox realized that it was on the verge of a crisis when Japanese companies were marketing photocopiers cheaper than it cost Xerox to manufacture a similar product. It is claimed that by benchmarking Xerox against Japanese companies it was able to improve their market position and the company has used the technique ever since to promote continuous improvement. Yet again another strong advocate of benchmarking is the automotive industry who successfully employed the technique to reduce manufacturing faults. Four types of benchmarking can broadly be defined: internal, competitive, functional and generic (Lema and Price, 1995). However Carr and Winch (1998) while regarding these categories as important suggest that a more useful distinction in terms of methodology is that of output and process benchmarking.

Interestingly (Winch, 1996) discovered that sometimes the results from a benchmarking exercise could be surprising, as illustrated in Table 6.4, which shows the performance of the Channel Tunnel project in relative to other mega projects throughout the world using benchmarks established by the RAND corporation. The results are surprising because the Channel Tunnel is regularly cited as an example of just how bad UK construction industry is at delivering prestige projects. By contrast the Winch benchmarking exercise demonstrated that the Channel Tunnel project fared better than average when measured against a range of performance criteria.

Since the Winch study in 1996 there has been a tailing off of traffic using the Channel Tunnel and it is now obvious that projections of growth of users was grossly optimistic.

Table 6.4 Mega project performance

Performance criterion	Mega projects average	Channel Tunnel
Budget increase	88%	69%
Programme overrun	17%	14.2%
Conformance overrun	53% performance not up to expectations	As expected
Operational profitability	72% unprofitable	Operationally profitable but overwhelmed by financial charges

Source: Winch (1996).

Measuring performance

Through the implementation of performance measures (what to measure) and selection of measuring tools (how to measure) an organization or a market sector communicates to the outside world and clients the priorities, objectives and values that the organization or market sector aspires to. Therefore the selection of appropriate measurement parameters and procedures is very important to the integrity of the system.

It is now important to distinguish between benchmarks and benchmarking. It is true to say that most organizations that participate in the production of KPIs for the Construction Best Practice Programme (CBPP) has to date produced benchmarks. Since the later 1990s there has been a widespread government backed campaign to introduce benchmarking into the construction industry with the use of so-called KPIs. The objectives of the benchmarking as defined by the Office of Government Commerce are illustrated in Figure 6.1. Benchmarks provide an indication of position relative to what is considered optimum practice and hence indicate a goal to be obtained, but while useful for getting a general idea of areas requiring performance improvement, they provide no indication of the mechanisms by which increased performance may be brought about. Basically it tells us that we are underperforming but it does not give us the basis for the underperformance. The production of KPIs which has been the focus of construction industry initiatives to date therefore has been concentrated on output benchmarks. A much more beneficial approach to

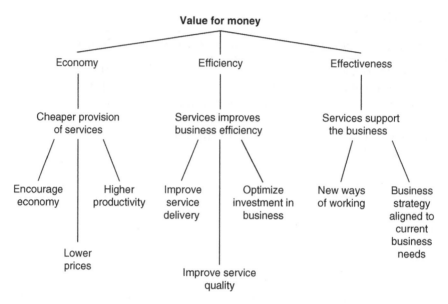

Figure 6.1 Value for money (source: OGC, April 2003).

measurement is process benchmarking described by Pickerell and Garnett (1997) '*as analysing why your current perform-ance is what it is, by examining the process your business goes through in comparison to other organizations that are doing better and then implementing the improvements to boost per-formance.*' The danger with the current enthusiasm in the con-struction industry for KPIs is that outputs will be measured and presented but processes will not be improved as the underlining causes will not be understood. According to Carr and Winch (1998) many recent benchmarking initiatives in the construction industry have shown that while the principles have been understood and there is much discussion about its potential 'no one is actually doing the real thing.' Bench-marking projects have tended to remain as strategic goals at the level of senior management.

The performance measures selected by the Construction Industry Best Practice Programme are:

- Project performance
  - Client satisfaction: product
  - Client satisfaction: service
  - Defects
  - Predictability: cost

- – Predictability: time
- – Construction: cost
- – Construction: time
- Company performance
  - – Profitability
  - – Productivity
  - – Safety
  - – Respect for people
  - – Environmental

and are measured at five key stages throughout the lifetime of a project. The measurement tools range from crude scoring on a 1–10 basis to the number of reportable accidents per 100,000 employees. For benchmarking purposes the construction industry is broken down into sectors such as public housing and repair and maintenance.

The balanced scorecard

Robert Kaplan and David Norton at Harvard Business School developed a new approach to performance measurement in the early 1990s. They named the system the 'balanced scorecard' (BSC). Recognizing some of the weaknesses and vagueness of previous management approaches, the BSC approach provides a clear prescription as to what organizations should measure in order to balance the financial perspective. The BSC is not just a measurement system, it is also a management system that enables organizations to clarify their vision and strategy and translate them into action. The BSC methodology builds on some key concepts of previous management ideas such as total quality management (TQM), including customer defined quality, continuous improvement, employee empowerment, and primarily measurement-based management and feedback.

Kaplan and Norton describe the innovation of the BSC as follows:

> *The balanced scorecard retains traditional financial measures. But financial measures tell the story of past events, an adequate story for industrial age companies for which investments in long-term capabilities and customer relationships were not critical for success. These financial measures are inadequate, however, for guiding*

*and evaluating the journey that information age companies must
make to create future value through investment in customers, suppliers, employees, processes, technology, and innovation.*

The BSC approach can be applied equally to both the public
and private sectors, although objectives in a public sector
organization will tend to differ, for example financial considerations in the public sector will seldom be the primary objective
for business systems. The BSC suggests that an organization
is viewed from four perspectives, as illustrated in Figure 6.2
and to develop metrics, collect data and analyse it relative to
each of these perspectives.

1. The financial perspective

Kaplan and Norton do not disregard the traditional need for
financial data. Timely and accurate funding data will always
be a priority, and managers will do whatever necessary to

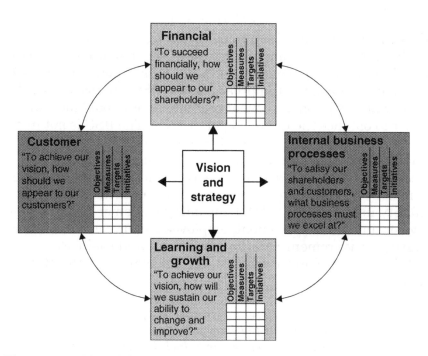

Figure 6.2 The BSC (source: Kaplan and Norton, *Harvard
Business Review*).

provide it. In fact, often there is more than enough handling and processing of financial data. With the implementation of a corporate database, it is hoped that more of the processing can be centralized and automated. But the point is that the current emphasis on financials leads to the 'unbalanced' situation with regard to other perspectives. There is perhaps a need to include additional financial-related data, such as risk assessment and cost-benefit data, in this category. Financial metrics could include: cost:spend ratio.

2. *The business process perspective*

This perspective refers to internal business processes. Metrics based on this perspective allow the managers to know how well their business is running, and whether its products and services conform to customer requirements (the mission). These metrics have to be carefully designed by those who know these processes most intimately; with our unique missions these are not something that can be developed by outside consultants.

In addition to the strategic management process, two kinds of business processes may be identified: (a) mission-oriented processes and (b) support processes. Mission-oriented processes are the special functions of government offices, and many unique problems are encountered in these processes. The support processes are more repetitive in nature, and hence easier to measure and benchmark using generic metrics. Metrics could include: percentage of actions utilizing electronic commerce, achievement of goals relating to, sustainability, socio-economic factors, etc.

3. *The customer perspective*

Recent management philosophy has shown an increasing realization of the importance of customer focus and customer satisfaction in any business. These are leading indicators: if customers are not satisfied, they will eventually find other suppliers that will meet their needs. Poor performance from this perspective is thus a leading indicator of future decline, even though the current financial picture may look good.

In developing metrics for satisfaction, customers should be analysed in terms of kinds of customers and the kinds of processes for which we are providing a product or service to those customer groups. Metrics could include: customer satisfaction with delivery, quality, responsiveness and cooperation.

4. *The learning and growth perspective*

This perspective includes employee training and corporate cultural attitudes related to both individual and corporate self-improvement. In a knowledge-worker organization, *people* – the only repository of knowledge – are the main resource. In the current climate of rapid technological change, it is becoming necessary for knowledge workers to be in a continuous learning mode. Government agencies often find themselves unable to hire new technical workers and at the same time is showing a decline in training of existing employees. This is a leading indicator of 'brain drain' that must be reversed. Metrics can be put into place to guide managers in focusing training funds where they can help the most. In any case, learning and growth constitute the essential foundation for success of any knowledge-worker organization. Metrics could include percentage of employees satisfied with management, work environment, health and safety. Also the number of employees with recognized training and qualifications.

Metrics

You cannot improve what you cannot measure. Often used interchangeably with measurements. However, it is helpful to separate these definitions. Metrics are the various parameters or ways of looking at a process that is to be measured. Metrics define *what* is to be measured. Some metrics are specialized, so they cannot be directly benchmarked or interpreted outside a mission-specific business unit, other measures will be generic. It is true that defining metrics is time consuming and has to be done by managers in their respective mission units. But once they are defined, they will not change very often. And some of the metrics are generic across all units, such as cycle time, customer satisfaction, employee attitudes, etc. Also, software tools are available to assist in this task. Metrics must be developed based on the priorities of the strategic plan, which provides the key business drivers and criteria for metrics managers most desire to watch. Processes are then designed to collect informa-tion relevant to these metrics and reduce it to numerical form for storage, display and analysis. Decision makers examine the outcomes of various measured processes and strategies and track the results to guide the company and provide feedback.

The value of metrics is in their ability to provide a factual basis for defining:

- Strategic feedback to show the present status of the organ-ization from many perspectives for decision makers.
- Diagnostic feedback into various processes to guide improve-ments on a continuous basis.
- Trends in performance over time as the metrics are tracked.
- Feedback around the measurement methods themselves, and which metrics should be tracked.
- Quantitative inputs to forecasting methods and models for decision support systems.

The goal of making measurements is to permit managers to see their company more clearly – from many perspectives – and hence to make wiser long-term decisions. The Baldrige Criteria (1997) booklet reiterates this concept of fact-based management:

Modern businesses depend upon measurement and analysis of performance. Measurements must derive from the company's strategy and provide critical data and information about key processes, outputs and results. Data and information needed for performance measurement and improvement are of many types, including: customer, product and service performance, operations, market, competitive comparisons, supplier, employee-related, and cost and financial. Analysis entails using data to determine trends, projections, and cause and effect – that might not be evi-dent without analysis. Data and analysis support a variety of company purposes, such as planning, reviewing company perform-ance, improving operations, and comparing company performance with competitors' or with 'best practices' benchmarks.

A major consideration in performance improvement involves the creation and use of performance measures or indicators. Perfor-mance measures or indicators are measurable characteristics of products, services, processes, and operations the company uses to track and improve performance. The measures or indicators should be selected to best represent the factors that lead to improved customer, operational, and financial performance. A comprehensive set of measures or indicators tied to customer and/or company performance requirements represents a clear basis for aligning all activities with the company's goals. Through the analysis of data from the tracking processes, the measures or indicators themselves may be evaluated and changed to better support such goals.

Generally as with any management tool, the results obtained either through the adoption of KPIs or BSC techniques must be taken seriously and must be seen to be adopted. Understanding what a particular result really means is important in determining whether or not it is useful to the organization. Data by itself is not useful information, but it can be when viewed from the context of organizational objectives, environmental conditions and other factors. Proper analysis is imperative in determining whether or not performance indicators are effective, and results are contributing to organizational objectives.

National Health Service ProCure 21 pre-qualification and final selection process

The pre-qualification forms the short-listing process for the selection of 12 candidates to proceed to the final selection. The build-up of the final selection total score is set out in Table 6.5.

Evaluation

The selection is based upon a quality/price score – the proportion is *70%* quality/*30%* price. The NHS is responsible for ensuring the optimum combination of performance capability and of delivering whole life cost efficiencies from the NHS ProCure 21 initiative.

The price scoring of *30%* is based upon the economic test. The *70%* quality score is subdivided in the following way:

- *35%* is allocated to the area of management ability and application of the principles outlined in Sir John Egan's report 'Rethinking Construction' and HM Treasury's document 'Achieving Excellence'. These principles are addressed under 'Egan vision/Achieving Excellence';

Table 6.5 NHS ProCure 21 pre-qualification

	Short-listing		Final selection process		
	PQQ questionnaire		**Visits**	**Economic test and financial review**	**Interview**
Process	**Part A Section 1 issues** General and financial information Detailed technical ability Experience	**Part B Section 2 issues** Best practice Egan vision/ achieving excellence	Site and office visits to evaluate systems and processes	Cost models	Standard interview questions and final clarification
Numbers	**Top 20**	**Top 12 proceed to FSP**	**Top five (maximum) proceed to framework agreements**		
% of final selection total quality/ price score	0	35%*	15%	30%	20%

* The Part B Section 2 PQQ score will constitute up to 35% of the final score and is considered as part of the final selection for those moving forward to FSP.

- *15%* is allocated to the ability of candidates' processes and systems to deliver NHS ProCure 21 objectives (site visits);
- *20%* is allocated to demonstrating integration and performance capability of candidates and their supply chain (interviews).

The best candidates average scores will form two short-lists of 12 candidates for the PFI and publicly funded frameworks.

Site visit programme

The final selection process (FSP) involves the completion of the cost models, the site visits and the interviews. This guidance outlines the site visit process for principal supply chain partner (PSCP) candidates. The maximum percentage score available from this process is 15% of the overall FSP score.

The weighting assigned to each of the component parts of the selection process is as follows:

- PQQ: 35%
- Site visits: 15%
- Cost models: 30%
- Interview: 20%

The site visits should enable evaluators to observe candidates interfacing with their primary supply chain members (PSCMs) and clients. Evaluators are to assess how effectively and efficiently projects are managed on a day-to-day basis. The candidate should provide tangible evidence demonstrating how they deliver complex, high quality projects on time and within budget. Examples of some of the major issues the evaluators will be looking for evidence of are as follows.

- Dynamic and successful team working, including active client and end-user involvement through partnering.
- Increased value for money for clients through whole life costs using innovative management and construction practices.
- Productive and cohesive interfaces with clients.
- Effective supply chain and resource management strategies and practices that minimize waste and maximize value.
- Efficient and effective site and contract management, organization and processes.
- Processes used to capture continuous improvements and how this information is utilized.
- Proven expertise in the provision of complex health care facilities.
- Sound environmental management strategies and techniques.

Candidates are required to:

- Brief the evaluators on these and any other issues which the candidates consider to be relevant.
- Provide copies of any relevant company policies and guidance together with an index of documents supplied. These should demonstrate clearly how these policies are implemented practically and the resultant tangible benefits to the end-user.

Format of the site visits

Each candidate will conduct a two-day site visit. The candidate must arrange a suitable timetable and determine the precise details of the programme within the following guidelines:

- The site visits take place over two consecutive days.
- The site visits which should include a complex health care project currently under construction must include the opportunity for evaluators to visit at least one completed and operational complex site. The evaluators will want to walk around the site and to talk openly to whomever they wish. They will need a guide to facilitate introductions and to comply with any health and safety requirements.
- Evaluators will need to sit as unaccompanied observers in a typical site meeting. The candidate will be required only to facilitate attendance, take the evaluators to the meeting, and to introduce them to the meeting chair. The evaluators will explain the purpose of their visit to the meeting if required.
- The site visit must provide evaluators with the opportunity to meet client representatives. These should be from a complex health care project completed by the candidate.
- The site visit should demonstrate the candidate's commitment, skills, knowledge and experience of some element of collaborative procurement processes. Example areas may include partnering, value management, risk management, whole life costs, etc.

Observers may accompany the evaluators from time to time. In such cases, the role of the observers will be simply to ensure that the evaluation process is working satisfactorily. They will take no part in the assessment process itself.

Interview objectives

The interview process gives candidates an opportunity to demonstrate what they can bring to the NHS ProCure 21 initiative. The candidates should demonstrate their proposals for delivering cost-effective, high quality, complex health care projects on time and within budget.

Examples of some of the major areas the interviewers want to assess:

- The technical skills and capabilities of PSCMs responsible for delivering NHS ProCure 21 projects. This includes the ability to work with clients and end-users in a partnering environment to maximize value for money.
- How the candidate selects and effectively partners with its supply chain. This includes how teamwork, continuous improvement, performance, client needs and culture are addressed.
- How a clear understanding of the specific needs and problems of health service clients and end-users are established. Also, how this information is taken into consideration during the development and delivery of complex projects.
- The candidates' understanding of the relationship between design, staff and patient wellbeing, and efficiency when operating complex healthcare facilities. How this understanding generates and implements considerations for whole life costing and value for money.
- How innovations and continuous improvements are implemented while meeting the need for completion, to high standards within tight time scales. Also, how difficulties are identified and successfully overcome when encouraging this approach throughout the entire supply chain.
- The candidates' understanding of the ECC form of contract using target costs and shared incentives.
- How the candidates successfully develop new approaches to old problems through successful partnering with clients and end-users. This should include maximizing overall value for money.

Conclusion

The Byatt Report, discussed in Chapter 9 attempts to extend the ethos of procurement on the basis of a balance of value for money and quality at local government level. It is clear that fewer and fewer procurement decisions, whether in the public or private sectors will, in the future, be taken on the basis of cheapest is best. As the basis on which construction projects are procured swings more and more towards non-adversarial long-term partnering with pre-selected main contractors and supply chains, then the basis for selecting partners and evaluating bids must also move away from the 'cheapest is best' philosophy to a more comprehensive and rigorous approach.

Bibliography

Carr B. and Winch W. (1998). *Construction Benchmarking: An International Perspective.* Engineering and Physical Research Council.

Flyvbjerg B. *et al.* (2003). *Megaprojects and Risk: An Anatomy of Ambition.* Cambridge.

Leibfried K.H.J. and McNair C.J. (1994). *Benchmarking: A Tool for Continuous Improvement.* Harper Collins.

Lema N.M. and Price A.D.F. (1995). Benchmarking performance improvement towards competitive advantage. *Journal of Management in Engineering.*

National Institute of Standards and Technology (1999). *Malcolm Baldrige National Quality Award 1999 Criteria for Performance Excellence,* Gaithersburg, Maryland, National Institute of Standards and Technology.

Websites

www.balancedscorecard.org/
www.quality.nist.gov/

7

Procurement – case studies

Introduction

This chapter includes two procurement case studies. The first is from the energy sector and is concerned with supply chain management (SCM) as well as procurement using reverse auctions (RAs), a practice that has been viewed with suspicion in some circles. The second case study is an example of an award winning procurement strategy; fairness, unity, seamless (ness), initiative, openness and no blame (FUSION), developed by Kevin Thomas, Director R&D Worldwide Strategic Planning, GlaxoSmithKline.

What is an online auction?

An online auction is an Internet based activity, which is used to negotiate prices for buying or selling direct materials, capital or services.

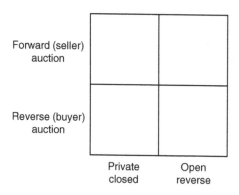

Online auctions can be used to sell – these are called Forward (or seller) auctions and closely resemble the activity on web sites such as e-bay. The highest bidder wins. However, most companies are starting to use RAs where purchasers seek market pricing inviting suppliers to compete for business on an online event. Auctions can either be private/closed where there are typically few bidders who have no visibility of each other's bids, or open, where a greater number of participants are invited. In this case participants have visibility of either their rank or the bidding itself. When used, the technique can replace the conventional methods of calling for sealed paper tenders or face to face negotiations. There now follows a case study of RA.

Case study 1: The use of Transformation Reverse Auctions in the BP UKCS SCM strategy

Rory Lamont, Supply Chain Specialist, BP
This case will firstly outline BP's United Kingdom Continental Shelf (UKCS) SCM Strategy. Second, it will present how BP makes use of Transformation Reverse Auctions (TRAs). As part of this, RAs will be discussed. Finally, a comparison between the traditional tender process, RAs and TRAs will be completed.

BP's UKCS SCM strategy

Within BP, SCM is thought of as the process by which BP decides how best to place its demand for goods and services into the external marketplace. It requires detailed analysis of demand levels and patterns and it requires an in-depth understanding of the marketplace and how it is contested. The process encompasses the first and second tiers of the supply base and covers the full range of activities from strategy through to contractor performance management.

BP is the leading producer of oil and gas in the UKCS, accounting for approximately 20% of the region's total production. This commitment to the North Sea represents an investment of approximately $2,000,000 million per year and demands a clear and efficient SCM strategy to ensure continued industry competitiveness. The development and maintenance of successful supply chain relationships is key in delivering world-class health, safety and environmental performance, efficient and cost-effective operations. It also allows BP to access new technologies available in the marketplace.

The global upstream SCM strategy within BP is encapsulated by the acronym STARS. STARS is the foundation that drives and shapes BP's supply chain tactics in the global upstream arena and stands for:

- *Sector strategies*: Maximizing value in key market sectors through a deep understanding of market and business drivers while ensuring security of supply.
- *Transparency*: Clarity of costs and performance.
- *Aggregation*: Leverage of spend across business units and streams.
- *Relationships*: Distinctive relationships based on performance, market reality and technology development.
- *SCM*: Clearly understanding BP's role in the supply chain.

STARS leads to the development of distinct sourcing strategies for each particular commodity or service sectors. It is BP segments that spend into sectors which reflect the structure of the marketplace. A sector team is appointed to manage the development of an appropriate SCM strategy, stretching from contract development and award to performance management. These teams are cross functional and include technical experts, operating staff, a safety expert and an SCM specialist. The SCM specialist is tasked specifically with managing the contractor relationship.

BP's global upstream SCM strategy STARS, has its roots in the current UKCS SCM strategy. The UKCS strategy was put in place in 1998 and is centred on four underlying principles that deliver value and growth. These principles are:

- *Manage the supply chain*: Maintain the clarity in the role that BP plays in the SCM process. This is about being an informed buyer and using detailed market information to understand and influence performance.
- *Utilize federal contracts*: Increase the use of regional agreements when appropriate across business units. This is to encourage long-term relationships, leverage volume, achieve appropriate dependencies and promote uniform behaviour across BP's UKCS operations.

- *Performance and cost transparency*: Put in place consistent cost and performance measures with targets and visible reporting. This is to encourage continuous improvement.
- *Access to new technologies*: Improved access to new technologies while encouraging, measuring and expecting innovation.

Using pre-1998 supplier and internal feedback, it is clear that prior to the implementation of the current principles, BP had a very fragmented approach to the marketplace in the UKCS. There was no clarity in BP's role in the supply chain and many SMEs felt excluded. There were no standard performance measurements, nor was there a process to encourage innovation.

Based on a review in late 2000, the results showed that BP's UKCS SCM performance had moved closer to that of a world-class organization and has delivered nine significant benefits:

1. Improved communication, both internal and external.
2. Standardized contract frameworks enabling consistent application of HSE standards.
3. Increased access to contractor management performance and cost data.
4. Increased focus on innovation and access to new technologies.
5. Clarity of BP's role in the supply chain.
6. Improved direct relationships with more SMEs.
7. Improved access for suppliers into the organization.
8. Increased understanding of the SCM principles.
9. Development of appropriate e-procurement tools.

Looking forward, BP faces significant challenges in the UKCS business environment. A reduction in average field size combined with the increasing depth at which finds are being made has pushed the boundaries of existing technologies. This in turn has placed great demands on the supply chain in terms of innovation and speed to market. It is therefore absolutely essential that BP achieve world-class performance from its first and second tier suppliers and contractors within the region.

Reverse auctions within the BP UKCS SCM strategy

BP group first trialled RAs in 1999. By the end of 2000, BP's upstream organization had completed $260 million worth of auctions for goods and services ranging from IT help desk services to seismic surveys. Several of these initial events took place in the UKCS. Since that time, BP has awarded well in excess of $1 billion worth of business through the RAs and the RA tool has become part of the standard sourcing tool kit of each supply chain specialist, along with traditional tenders and negotiations.

In 2000, a review of the trial events was carried out. It identified three characteristics that need to be present to have a successful RA: the purchase must be clearly defined; the market must be well contested and BP needs to have existing knowledge of the supply base. These three factors are interdependent and together form the basis for an auction that delivers final prices as close as possible to the 'true' current market prices. For the buyer and supplier, a clearly defined scope of work is essential. Without this it becomes very difficult accurately to bid for the work. Contestation, that is three or more suppliers within the market willing to bid for the work, is another pre-requisite. Without this, there is no incentive for suppliers to reconsider their proposals. Finally, BP's knowledge of the supply base ensures that the most suitable suppliers participate in the events. Since this review, the above three factors have been adopted by BP's SCM specialists as the criteria for whether or not to use a RA.

The review highlighted the focus on price as being a weakness in the RA process. This was considered to be the case because BP's policy had always been to evaluate supplier proposals on a range of criteria, not uniquely on price. These additional criteria, such as safety factors or technical ability, were not reflected in the ranking that each supplier received during the event. Therefore, the suppliers could not see how their proposals compared to their competitors overall.

BP decided that if suppliers submitted proposals excluding price information before the auction, these proposals could be evaluated beforehand and the resulting scores built into the auction tool. Therefore, when a supplier entered their price, the application already had the information needed to

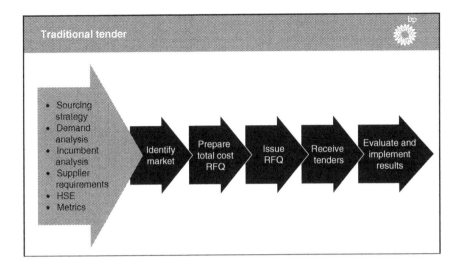

Figure 7.1 BP case study – traditional tender.

complete the evaluation process and present this on a graph. This type of auction, where a supplier's scores and pricing information were evaluated within the application itself is known as a TRA.

Comparison: tendering process and TRA process

It is clear that the process followed in executing a TRA differs from that of a traditional tender process. As can be seen from Figure 7.1, typically there are five stages in BP's tendering process, depicted by the black arrows. The precursors to these stages can be seen in the large arrow on the left.

Figure 7.2 represents the process that is followed when a TRA is used; the small arrows representing stages specific to the TRA process. The large arrow is a step generic to both the tender and the auction process. As part of setting the sourcing strategy, the supply chain specialist uses the three criteria identified in the 2000 review to decide whether or not an auction is suitable. If the criteria are met, then the recommendation will be made to the business to proceed, using a TRA to source the goods or services. Along with the pre-qualification of potential suppliers, the supply chain specialist must also develop a lotting strategy for the auction.

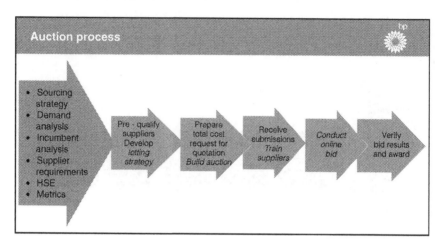

Figure 7.2 Auction process.

An auction may consist of several lots, each of which may cover all or part of the scope of work. The lotting strategy is dependent on two factors: the capabilities of the supply base and the sourcing strategy developed. If the sourcing strategy for is to buy the entire requirement from one supplier, there may only be one lot covering the whole scope of the work. If BP is interested in unbundling the work, then individual lots will be constructed that, when amalgamated, cover the required scope. In this scenario, it is conceivable that not all suppliers invited to participate will bid in all lots. In both cases, the choice of which lots to participate in remains the supplier's.

Once the total cost request for quotation (RFQ) has been issued, the SCM specialist must create the auction on the auction software. This involves logging into an externally hosted application and entering the details of the event, such as start and finish time, lot details, related documents, the name and e-mail addresses of the suppliers and the transformation factors (see section Calculating Transformation factors: A worked example). The application then issues the selected suppliers with the URL for the site and individual passwords. At this point, each supplier has the opportunity to review the details of the event and accept or decline to participate in each lot.

Before each auction, training takes place in the form of a fictional auction. This allows suppliers to understand the login process and how the bidding functionality works. BP also talks to suppliers about the need to have a bidding strategy developed beforehand. As a minimum, BP recommends calculating a lowest bid for each lot. The over-riding message that BP leaves suppliers with is, 'Be prepared!'

On the day of the bid, a team of BP employees conducts the auction. This team consists of a buyer, often the SCM specialist involved in the sourcing process, an auction administrator and surrogate bidders. Before the event, the buyer will check the application to ensure that all details of the auction are still correct. They will also check which suppliers have agreed to participate and confirm with all invited suppliers that they are not experiencing any difficulties with the software. The auction administrator will check with BP's IT change management team that the BP network infrastructure is stable and make sure that one surrogate bidder is available for every two participating suppliers.

The surrogate bidders are present in case suppliers experience technical difficulties during the event. In such a situation, they are instructed by the buyer to contact the suppliers by telephone and, at the supplier's instruction, place proposals on their behalf. Surrogate bidders have no access to the buyer or administrator and must adhere to a pre-prepared script. To further ensure impartiality, they only have access to the same information and functionality as the supplier on whose behalf they are acting.

The buyers are responsible for ensuring the integrity of the auction; including ensuring that malicious supplier behaviour cannot disrupt the market. Malicious behaviour could take the form of one supplier trying deliberately to lead the market price down, by taking the lead with a very low bid, waiting a short period to allow others to overtake their lead and then asking for the bid to be removed. BP has developed a set of market rules by which all RA are conducted (See Table 7.1). As well as giving the buyer guidance, these rules also define the standards by which auctions are conducted so a supplier taking part in an auction held for the UKCS will participate under the same rules as a one participating in Indonesia.

Within BP, a global auction team consisting of four Auction Champions supports these teams. The Champions are also charged with ensuring that the market rules and supplier training are consistently applied and updated.

The process impact of using TRA on the buying organization has been previously outlined in Figure 7.2. The white tasks, which are completed by the buyers, are in addition to those

			bp
Table 7.1　BP case study – Market rules			

Issue	Steps			
	1	*2*	*3*	*4*
Erroneous bid (market leading or not) reported to Buyer *in first 90% of a lot*	*Buyer* instructs admin to pause event	*Buyer* confirms supplier's ID *Buyer* instructs admin to remove bid	*Buyer* tells supplier that on third erroneous bid, will be removed from event	*Admin* resumes event
Erroneous market leading bid reported to Buyer *in last 10% of a lot*	*Buyer* instructs admin to pause event	*Buyer* confirms supplier's ID	*Buyer offers* to remove erroneous bid any bids placed in response to erroneous bid	*Admin* resumes event
Multiple erroneous bids	*Buyer* monitors all erroneous bidding	*Buyer* issues warning on each bid	*Buyer* instructs admin to remove supplier on third erroneous bid	
No activity from a supplier	*Buyer* will construct lots to include starting gate	*Suppliers* only contacted if technical issue detected		
Supplier loses connection	*Buyer* instructs admin to pause event and issue broadcast message with reason for pause	*Supplier* tries to reconnect. (Max 10 mins)	After 10 minutes, *Buyer* gives supplier a surrogate bidder's number (*See surrogate bidding process*)	*Buyer* instructs *admin* to resume event once surrogate is ready to act
Admin receives supplier call re technical issue	*Admin* passes details of call to buyer	*Buyer* takes action in accordance with market rules		

carried out in the traditional tender process. The principal change that takes place is the moving of the proposal evaluation to earlier in the sourcing process. This allows the scores to be built into the application before the auction is executed. A final check of the auction results is carried out post-event and in the majority of cases; the contract award will be given to the lowest transformed bid, *ceteris paribus*.

The extra time it takes to complete an auction in this way has been estimated internally at one man-day. Often the tasks of building the auction and training suppliers take place in parallel with the proposal evaluation, thus the impact on any sourcing project timeline is minimal. However, the process does call on coaching and training skills not always associated with supply chain staff. To help the SCM specialist with this, BP has developed standard training materials for use with suppliers.

Calculating Transformation Factors: A worked example

The transformation factors entered into the RA application can be any combination of multiplying factors or additional factors. Typically, they adhere to the form:

Transformed bid = (Raw bid) × Multiplier + Adder

Transformation is no different from what BP already does in a traditional tender evaluation process, except the evaluation takes place pre-auction.

Let us look at an example – provision of pumps, including storage before use. The scope of work has been fixed and four potential suppliers have been identified, including the current supplier; the scene is set for a TRA.

To make use of a transformation auction, the supply chain specialist defines the evaluation factors and weightings prior to the RFQ being sent out to the suppliers. The information is then collected from each supplier in order to complete the assessment.

The next step is to convert these scores into multipliers that can be applied to the relevant supplier's proposals by the software. These factors express each supplier's evaluation relative to the most attractive suppliers.

Adders can be useful when there are switching costs associated with moving from the incumbent to another supplier. To continue the example, let us think about moving costs for the pumps currently being held by the existing supplier. Imagine that Supplier 1 is the incumbent; it is highly likely that there will be a cost incurred by BP to move the stock to a new supplier and update their inventory records in the event of switching the business. If this information is solicited from suppliers before the auction, it can be built into the software. Multipliers and adders have now been developed for each supplier taking into account their performance and other associated costs. In this example the factors, weightings and scores can be seen in Table 7.2. The effect of the transformation factor on Supplier 1 would be to increase any bid by 2.8%. On Supplier 2 it would be to increase their proposals by 3.5% and add $150,000 and so on. This gives BP an instant view of the marketplace in terms of the total cost of acquisition for each supplier.

Conclusion

TRA form part of BP's supply chain specialists' tool kit. They are a sourcing process with their own inherent advantages and disadvantages for BP and suppliers – see Table 7.3.

Table 7.2 Supplier evaluation				bp
	Supplier 1	*Supplier 2*	*Supplier 3*	*Supplier 4*
Health and Safety (35%)	80/100	85/100	90/100	92/100
Technical (50%)	95/100	89/100	89/100	87/100
Environmental (15%)	70/100	75/100	80/100	85/100
Total	86	85.5	88	88.45
Conversion	88.45/86	88.45/85.5	88.45/88	88.45/88.45
Multiplier	1.028	1.035	1.005	1
Adder ($)	0	150,000	200,000	220,000
Transformation factor	1.028 (bid) + 0	1.035 (bid) + 150,000	1.005 (bid) + 200,000	1 (bid) + 220,000

Total = [{Health and Safety score × 35%} + {Technical score × 55%} + {Environmental score × 15%}].
Multiplier = Best Supplier TOTAL/Actual Supplier TOTAL.
Transformation factor = multiplier × raw bid + adder.

Party	Process		
	TT	*RA*	*TRA*
BP	+Accepted norm −Comparison between suppliers limited to evaluation of one price proposal per tender submission −Totally manual evaluation process.	+Time constraint encourages competition (Auction effect) +Auction effect reduces price variation between suppliers +Automated supplier evaluation against price −Auction graph reflects price only −Manual evaluation of other factors required after auction −Need adequately contested market −Potential supplier resistance	+Time constraint encourages competition (Auction effect) Auction effect reduces price variation between suppliers +One time manual evaluation of non-price elements done pre-auction, tool manipulates supplier's bids to reflect evaluation +Auction graph reflects evaluated supplier proposals +Minimal need for manual evaluation −Need adequately contested market −Potential Supplier resistance
Supplier	+Accepted norm −Limited to one price proposal per tender submission −Price proposal based on perception of place in market (bidding blind) −No way to gain market knowledge −Post-tender feedback can only be used in next tender exercise	+Feedback on position in market based on price +Ability to put real-time feedback to use in same tender exercise −Misperception that lowest bid wins	+Feedback on position in market based on all evaluated criteria +Ability to put real-time feedback to use in same tender exercise +*Ceteris paribus,* lowest bid wins

TT: Traditional Tender; RA: Reverse Auction; TR: Transformation Reverse Auction.

The main purpose of either type of RA is to automate and accelerate the evaluation of proposals. The advantage that this brings to both BP and suppliers is instant information feedback. While feedback is available in a traditional tender, it takes longer to acquire and suppliers can only use it to revise their aspects of their proposals for subsequent tenders.

In a RA the evaluation criteria are limited to price. A TRA evaluates and graphs suppliers' proposals according to the individual supplier evaluations entered into the tool by BP. This allows BP to see very clearly the most attractive supplier. In turn, the suppliers see their rank in the market. In either type of auction, the suppliers may use this information to revise actual proposals. The Transformation auction format offers the most comprehensive evaluation of proposals and excluding significant mitigating factors, the lowest transformed proposal wins the contract.

Criticism has been levied by suppliers that RAs are used to drive down market prices. TRA, however, create a virtual model of the market for the goods being sourced. In such a model market forces differentiate competing proposals. Any resulting price reductions are not attributable uniquely to the sourcing process but to a combination of this and the prevailing market dynamics.

Case study 2: FUSION, all together a better way – a case study in collaborative working

Kevin Thomas, Director R&D Worldwide Strategic Planning, GlaxoSmithKline

Growing awareness

When we are born we are totally dependent on others for our nourishment, protection and development. We are so unaware of the world that we do not even know we have feet, let alone toes. We have no sense of purposes, other than to survive, and little sense of self.

As we grow we become more aware. We discover our limbs and begin to make them work for us. We learn how to communicate to ask for what we want and when we want it. We begin to establish goals and develop strategies for achieving them. We begin the process of becoming independent.

This struggle for independence is a long one, some never fully achieve it. But most of us reach a point at which we are able to exercise free will in all our actions. However, as our awareness grows we begin to realize that we cannot achieve many of the things we desire on our own. We find that co-operating with others is often more productive and we begin to build our credibility account, depositing assistance for others and withdrawing favours for ourselves.

Though social interaction, education, business and leisure, many of us build attachments with individuals and groups, which grow so close that we become fully interdependent on one another. We trust each other so implicitly that we do not even bother to develop all the skills and collect all the knowledge that we need, since it is available through our partnerships. We realize this makes us stronger, wiser and able to achieve more, we also realize that this is the way others become successful.

Eventually, we begin to realize that all these interdependencies are interconnected. We share the same limited resources. We achieve most when we are all successful and we lose most when we all fail. Whatever you call this, perhaps highest level of awareness, it is rarely attained by individuals and is sorely missing in the world at large.

This growth cycle is not just about individuals; it also applies to communities, businesses, industries, etc. and it was this growing awareness which, following the merger in 1995, lead the newly formed Glaxo Wellcome to pilot an alternative approach to construction activity.

Called FUSION, this was a collaborative approach based on fundamental team values of fairness, unity, seamless (ness), initiative, openness and no blame. The approach initially evolved during the major redevelopment of one of the company's laboratory buildings at its Ware R & D Site in Hertfordshire, England, before being applied from inception on the restructuring of the Beckenham site in Kent, England. The Beckenham Project went on to win the 1999 *Contract Journal* single project partnering award, so what made this FUSION Project so special?

Let us start at the beginning, not only because it is the best place to start but also because this is where the most profound changes took place.

What's wrong with the existing model?

The traditional way of undertaking a project is to assemble a team of designers to consider the business need and to investigate the feasibility of a solution (Figure 7.3). This design team, normally supported by quantity surveyors, then develop the scheme, outline and detail designs which are used to procure the implementation team of contractors, suppliers, manufacturers, etc.

The isolated components of the implementation teams are then asked to tender against the design drawings. Usually this is the implementer's first sight of the drawings and in the limited time on offer, they make assumptions about the intent, misunderstand some of the content, make mistakes in translating their supply chain's proposals, offering alternatives which are not really equivalent, and often make flawed 'commercial' decisions on how and where profit will be made. They generally hold back their concerns about the detailed proposals as experience shows them that designers tend not to recommend appointment of people who criticize the design. Despite the best endeavours of the selection team, many of these problems cannot be identified in the appointment process. However, once the contracts are let, the implementation team comes together and voice their concerns, suggesting different solutions and pointing out the weaknesses and inconsistencies in the design drawing and specifications.

Why does all this happen? Quite simply because the concept of asking a group of people who don't do implementation to develop solutions for another group who do, but without consulting the implementers in the design process, is fundamentally flawed. It leads to lack of ownership and buy-in with the all too familiar consequences of misunderstanding, disagreement, conflict, abortive work, delay and in the worst case claims and liquidations. No wonder the outcomes are frequently suboptimal to say the least and in the extreme not really fit for the purpose for which they were originally intended.

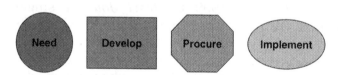

Figure 7.3 Traditional procurement process.

Redesign the process

The FUSION approach radically re-engineered this process. The starting point was still the identification of the business need. Initially this was the disposal of the over 40-hectare Beckenham R&D site after the functions housed there had been relocated to sites in Greenford, Stevenage and Ware. As the project developed it was clear that the biotechnology functions would be prohibitively expensive to re-provide elsewhere and so the specific need to consolidate biotechnology onto around 10 hectares of the site became the starting point for the restructuring project. A small master planning group was assembled to assist in developing the high level plans and indicative budget. Once this 'scale' work was completed an appropriate team with the necessary capabilities to handle the project at the scale identified was assembled.

This assembly was a progressive activity. First the space planners were selected and appointed. They joined the selection teams to appoint services and civil/structural engineers, who joined the team to appoint a builder, who joined to appoint a mechanical contractor, etc. In all, ten primary partners were appointed (not all interviewed by everyone, the expanding team agreeing who should be included in the selection panel). All the primary partners were contracted direct to Glaxo Wellcome, including a laboratory furniture manufacturer and a communications systems installer, as both were considered critical to the success of the project. Appointments were made on an open book basis, meaning payment would be made on the actual cost of goods and services provided, with an agreed mark-up for overheads and profit. Common commercials terms were applied which had no provision for set-off and liquidated damages set to zero.

There was no main or managing contractor appointed, instead a 'Principals Chamber' was formed comprising the ten primary partners together with Glaxo Wellcome. This chamber provided equal status for the principal members of the primary partners to agree the principles of the project. Chaired by the project sponsor, the Principals Chamber was assembled at the onset and its first activity was to agree who should do what. Roles were allocated to those best suited irrespective of which company they came from. The project team was assembled under a programme manager who was also a member of the Principals Chamber and the primary conduit to the project team.

Become a seamless team

The project could now begin in earnest. A quick start process was introduced with an off-site team building workshop. All the key members of the project team were assembled to gain an understanding of collaborative working as described in FUSION, to explore each other's roles and responsibilities and to reach common agreement on the goals and objectives. At this workshop the team agreed a charter, a project logo and a three-statement value proposition:

- *Scope*: A realistic solution to the business requirement
- *Programme*: Optimize time, maximize opportunity, realize success
- *Cost*: Input partnership, output value.

The project team was then ready to investigate the high level plan, the target budget and to develop a detailed scope and cost plan for the project. Now the radical differences of FUSION become clear. The team is free to decide how much time and effort is needed to be expended on design, including who is best to undertake that design and how much design is necessary before manufacture, assembly and installation should progress. There were no approved design drawings to be 'handed off' to be turned into working drawings by someone else, merely to apportion accountability (blame) at a future date. There were only agreed fabrication/installation drawings reflecting what was actually to be implemented, to be updated as work took place and become as installed drawings that actually reflected what happened. In this way design is able to proceed as a bow wave ahead of installation, with the team deciding what 'gateway' decisions are needed at what point in time and leaving details to be defined when they are required. With all the primary partners involved in the designs and decisions that affect them most, there is buy-in and ownership of the outcomes and most problems were identified and resolved while they were still on the drawing board (Figure 7.4).

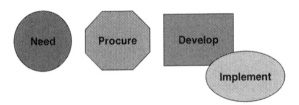

Figure 7.4 The FUSION process.

Of course this process was new and there were misunderstandings and differences of opinion, there were also problems which made their way onto the site in the guise of co-ordination clashes, delivery and availability issues, quality control and the inevitable spectre of changes and variations. The ways in which these were resolved go further to illustrate the radical changes delivered by FUSION.

Sometimes it's only the problems which highlight the success

Looking first at site problems. The traditional contractual models require people to identify problems and act on instructions received in response. In practice there is frequently a significant time delay following a tradesman advising their foreman that there is, say, a co-ordination problem. That tradesman is contractually bound to carry on the installation while the foreman advises the agent or contract manager who passes on the concern to the clerk of works who advises the design team, who then rush to find a resolution to be issued as an instruction so that the tradesman knows what to do to resolve the problem. In the meantime the demotivated tradesman has had to continue the installation, as have all the other trades impacted by that progress and very likely this has led to poor quality work being installed. When the instruction is eventually received, the extent of abortive work can be determined and this will be removed and replaced. However, sometimes solutions which retain some or all of this work are identified, this sounds better as it reduces the rework, but now poor quality work has been incorporated which will cause problems later on, possibly at commission or testing when remedial work is much more expensive to undertake.

At Beckenham, the site operatives were empowered. They were authorized to stop if they found a co-ordination clash of this type. They were also permitted to suggest alternatives and to call in the design team if they needed advice or if the change was outside their level of expertise, or required evaluation of the impact on other trade activities or design parameters. They were also empowered to seek other workfaces while the issues were being resolved or paid for standing by if there were no alternatives available. Consequently the amount of abortive work at Beckenham was dramatically reduced leading to significant savings in material and man-hours. Furthermore, the commitment of the workforce remained high and a steady stream of suggestions and ideas for improvements flowed from the workface.

Not that this happened by chance, every person joining the project was given an induction into the FUSION Values, was explained the project goals and objectives, what they meant and why they mattered to Glaxo Wellcome. This philosophy extended throughout the project team and everyone was encouraged to use their initiative. Occasionally people would make mistakes and these were rectified without penalty and treated as learning opportunities for the whole team, provided they were not as a result of neglect or naïve decisions for which people were expected to take responsibility.

A seamless team must include the customer

This engagement and freedom to act was only possible because the boundaries which normally fragment the team were not permitted on the FUSION Project. Everyone was considered part of the same team, encouraged to work across company boundaries and to challenge and question others to seek clarity and understanding and to offer support and advice. By understanding that success could only be achieved if everyone succeeded, a culture of proactively seeking and resolving problems before they become critical, could evolve. This was primarily a culture based on collective agreement, but of course it was supported by processes for escalating issues if agreement could not be reached.

Most escalation was contained within the project team, but occasionally the significance of the issues meant that resolution would be escalated to the Principals Chamber. This was particularly true of client changes. Variations which would not significantly compromise the budget or programme, were accepted and incorporated by the project team, whereas variations with significant impact were escalated. In the case of client changes there was an additional body formed to consider the effects of significant changes in scope. Called the 'Project Board' this body comprised senior representatives of the end user functions together with the project sponsor and programme manager (Figure 7.5).

To understand how these bodies interacted and how decision making could be effectively delegated without compromising the project budget it is necessary to understand the cost management process. Essentially it was the open nature of costing which was key to effective decision making. The budget was constructed as an elemental cost plan. In the early days many of these elements were unit rates based on previous experience, but as the project developed, the costs were developed to reflect the detailed decisions made and additional information available. Costs for all necessary services and activities were defined, including client related costs such as planning approvals and building regulations. Contingency was agreed and identified for all to see and once the general scope was defined and agreed contingency allowances were allocated to the Principals Chamber and the Project Board to deal with 'growth' (design development and implementation related errors and omissions) and scope changes respectively.

In this way every member of the team was able to understand the implications of changes in their area on the overall

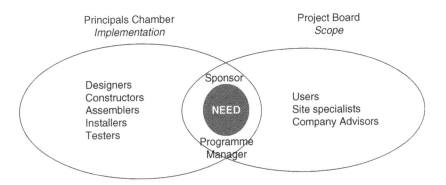

Figure 7.5 FUSION project organization.

cost plan and were able to make value based decisions. Furthermore, members were able to redistribute the budget to more appropriate areas as details were confirmed. The Project Board would review end user proposals for change, considering the cost, time and feasibility advice provided by the project team, sanctioned where necessary by the Principals Chamber. An end user sponsor would be required to present the case for scope change to the Project Board which, if agreed, would be drawn down from scope change contingency. This process led to buy-in to the budget from all parties and, not surprisingly, brought the project in both early and under budget.

Manage expectations

Another way of managing change is to recognize that many variations are due to miscommunication between users/specifiers and the project teams. This leads to differing levels of expectations which often translate into dissatisfaction with the end product no matter how well the project has been delivered. This was addressed at Beckenham by the creation of a Customer Care Team drawn from within the project team and tasked with ensuring that no end users were surprised by the facility on occupation.

A number of techniques were used by this group including, mock ups and sample displays, but by far the most effective was stopping the project activities on a regular basis to allow site familiarization visits. This enabled the end users to monitor progress on their element of the project and to raise issues and concerns. The Customer Care Team was allocated a limited budget to accommodate any changes they judged beneficial providing they could be accommodated without compromising the programme. This showed benefit on many levels. On one notable occasion a user visiting their area enquired what the yellow lines on the wall meant. When told it was marking for a conduit chase to a low level socket outlet as shown on the drawing she had signed off, she responded '*oh don't do that, I'm going to put some lockers there after you've gone*' demonstrating the folly of believing that the average person is able to visualize the messages contained in two-dimensional drawings. The chase was repositioned, saving the abortive cost of abandoning the outlet at a later date and rewiring to a newly installed outlet, satisfying the end user for the cost of a yellow chalk line.

When should handover happen?

After the first phased completion, the Customer Care Team introduced another innovation on the Beckenham Project. Handover is normally signified by the in-house maintenance team, fire, safety and end users producing snagging lists of outstanding defects identified prior to occupation. Until these lists are agreed handover does not take place and the area cannot be occupied. This is primarily a process of apportioning responsibility and limiting subsequent change. However, end users are suspicious of this process and tend to list issues 'just in case,' as it may be their only opportunity. Furthermore, people are not well accustomed to change and they will tend to question new provisions and list as missing things they used to have, whether or not they need them.

On the Beckenham Project the only handover documentation prior to occupation was to confirm that the area was safe to use and operable. The Customer Care Team was responsible for ensuring that there were no major omissions and the end users were not asked to confirm that the area could be accepted until they had used the facility for a couple of weeks and there were no outstanding issues. This led to snag-free handovers and delighted end users as the lack of pressure to agree to completion before occupation and the time to get used to the new facility, eliminated many of the previously identified issues. This also delivered benefit to the project team as the snag-free methodology was supported by a policy of non-retention of contract funds.

The Customer Care Team took additional responsibility for the training and support of the Maintenance Team, ensuring that all necessary documentation was provided. In this role they were supported by a commissioning manager appointed early in the project to review designs for operability and to ensure documentation was developed in parallel with project delivery.

Success to celebrate

The project was completed in phases and each phase was collectively celebrated as the close of a cycle of induction, training and ongoing support which typified the project. There were

plenty of opportunities for celebration, but just how successful
was the project? This can only be understood by looking at the
needs and how they were satisfied.

The primary need was to consolidate biotechnology functions
onto a corner of the Beckenham site releasing the remainder for
sale and redevelopment. The target budget for this consolidation was over £15 million. This required the 'unbuilding' of the
40+ hectares Beckenham site which, like most sites, had grown
'like topsy' with no consideration to future disposal (a good
lesson for us all to learn if we are serious about sustainability).
This means disentangling the services while maintaining the
biotechnology operations throughout the 22-month programme.

The disposal strategy was to play a big part in this rational-
ization as retention of part of the site as a science park or industrial or commercial business units, would all significantly
impact the services capacity and distribution methodology. This
disposal strategy changed regularly during the project, eventually ending up as a complete demolition and decontamination
for resale as a housing site, although this was understandably
resisted by the Local Authority, Bromley Council, who would
have preferred to retain more employment. Despite this constantly changing backdrop, the Biotech Formation Project
Teams, as they decided to be called, delivered 20 infrastructure
projects ranging from new security fence and protection systems
to the diversion of surface water drainage running directly
under the site. This was matched by 20 different facility projects, from restructuring existing laboratories and production
suites to constructing new office, boiler house and warehouse
facilities. The nature of many of these facility projects was also
changeable as the evolving Glaxo Wellcome Biotechnology
Division was subject to organizational change during the early
part of the programme leading to variation in the types of size of
functions to be housed on the Beckenham site.

With so much change, the programme was a recipe for disaster, yet it was enormously successful, with each of the individual facility and services phases being delivered to an agreed
sequence and schedule with very few sectional delays or over-
spends. This was because the various elements of the team
worked in complete harmony. The Project Board was allocated
only £80,000 for scope revisions, which they were entitled to
augment by delivery of scope reductions. The Principals
Chamber had the lion's share of the growth contingency, but at

7% of budget this was well below the level normally associated with a programme of this scale and uncertainty. As has already been stated, the key to this was the open nature of the cost plan, which meant at all times everyone could see how the various activities added together to deliver the target. Even the £15,000 given to the Customer Care Team to be spent on whatever additions they agreed, was visible to all parties and considered money well spent. This target did not overspend because in many ways it could not – it required everyone to agree to accommodate a change which could not be afforded, in order to produce an overspend. On the one occasion, nearing the end of the last phase, when an additional business need arose that all parties felt would be of real value to Glaxo Wellcome, the function promoting the need refused to allow it to proceed if it would undermine the successful conclusion to the programme.

At the end there was 1% remaining in the growth pot which was divided equally among the partners who agreed at onset to a share risk/reward formula. The formula offered a maximum benefit of £50,000 per partner on underspend and a maximum penalty of £10,000 per partner on overspend. However, on the next and last major FUSION Project for Glaxo Wellcome, the financial incentive was dropped. This was because all the partners felt that the 'guaranteed' nature of profit and the opportunities of being able to participate in such an enjoyable, rewarding and successful experience, for all levels of the organization involved, made the financial nature of the incentives trivial by comparison. Indeed in some companies they decided not to again face the dilemma of how to distribute the financial 'bonus' to the project staff, without undermining the relationships with other staff members not involved in such a 'pleasant' project. As one of the principals put it 'I've got people who have to put up with contractual behaviour all day long, asking why a colleague who has been working in a rose garden gets a pay out.'

Conclusion

For GlaxoSmithKline, the outcomes are far more than on time and under budget. The facilities and services delivered met and have continued to meet the evolving needs of the site, because the team understood, really understood, what biotechnology was about and how it might change in the coming years. End users were delighted with their facilities and felt really valued by the way they were involved and consulted. Many of the partners continued relationships with GlaxoSmithKline transporting the lessons learned to other sites. At the post programme review (another feature of FUSION Projects) the team compared the actual performance against their collective understanding of how it would have worked in a 'traditional' (we must start to call this 'historic') model. They concluded that the work was completed 37% earlier than the traditional model would have allowed and permitted 21% of additional value to be added, which in traditional contractual relationships would have been lost to abortive work, rework and additional management resources. All of which demonstrates the power of fully integrated, genuine collaborative working which truly is, 'All Together A Better Way.'

Note: Much of the learning from the Glaxo Wellcome Fusion Project has been incorporated into the Strategic Forum for Construction Toolkit on Integrating the UK Construction Industry. *Glaxo Wellcome merged in 2000 with SmithKline Beecham to form GlaxoSmithKline.*

Web sites

www.purchasing.com
www.fusion-approach.com/

8

Procurement – studies from other countries

Introduction

This chapter examines the procurement methods used by countries other than the UK. The reasons behind the development of national procurement practice lie in a number of areas including; historical, legal, social and financial. The aim of this chapter is not a detailed historical account but rather to analyse the drivers and the characteristics of the development of construction procurement.

Systems compared

For the purposes of carrying out a comparison of different approaches to construction procurement countries have been grouped as follows:

- France and Belgium being examples of countries with codified constitutions and legal systems.
- Australia being an example of a country with a federal system.

Codified constitutions and legal systems

The following statement appears in Chapter 8 of *Building Procurement Systems – A Client's Guide* by J. Franks, 1998:

> There is, then, a rag-bag of alternative systems for building procurement operating in Europe, with which most, if not all, of the British industry is familiar. There is ample evidence that in

property development the United Kingdom is probably ahead of other countries in Europe.

In fact the so-called 'rag bag of alternative systems' have developed, not in any haphazard way but in response to a variety of circumstances. Within Europe the factors that appear to have most impact on construction procurement development are as follows.

The systems of civil law

That is to say, does a country or state have a system based on civil code or common law, or construction law. It would appear that countries that have unfussy contractual relationships, unlike the UK that in large parts remains wedded to the Joint Contracts Tribunal Standard Forms, have less confrontational systems. No other EU state, for example, has adopted the UK model of lump sum contracts, which presumes that delays and alterations are inevitable and that the client shall pay for them, with the blame being laid at a variety of sources.

The availability of project latent defects insurance

As discussed in Chapter 2, latent defects insurance is taken out in only approximately 10% of all UK construction projects and to date the UK construction industry appears to regard the costs associated with this form of cover unattractive. France and Belgium both operate project insurance systems that will be discussed in more detail shortly. As will become apparent, the major advantage of project/latent defects insurance, from the procurement perspective is, that once in place, it allows the contractor to innovate and value engineer architectural concepts to deliver added value. In addition there is no doubt also, that the presence of project insurance helps to engender greater collaboration and interaction between parties during the procurement stage.

The status and education of construction professionals and contractors

Until their merger in 1982 the UK had two principal organizations for quantity surveyors, the Royal Institution of Chartered

Surveyors (RICS) and the Institute of Quantity Surveyors (IQS). The main reason for the existence of two similar organizations was that historically, Chartered Quantity Surveyors were not permitted to work for building contractors, if they did they had to resign as a member of the RICS and hence the IQS, an organization primarily for contractors' surveyors was founded. Presumably the logic behind this RICS ruling was that members of its organization, who had 'gone over to the dark side' were working within companies of which the Institution inherently disapproved. Prior to the merger of the RICS and the IQS in 1982, the RICS rescinded this ruling, but nevertheless this mind set speaks volumes about the attitudes in the UK construction industry, not to mention its professional institutions at this time. By contrast, in France, it has never been frowned on to work for a contractor and consequently highly trained professionals have always worked within contractors' organizations. The UK is virtually unique in having large professional practices of construction consultants with hundreds of staff on their payroll. In most other countries in the world, professional practices are small and typically in the case of project design, concentrate on the concepts, rather than detail and execution which are retained for the contractors. From the procurement perspective the advantages of this approach once again is the integration of the design and procurement, as well as placing the responsibility for detailed design and execution with the contractor, thereby assuring buildability (Table 8.1).

Generally speaking the UK has more varieties of procurement strategies than most other European countries, indeed it is not Europe that could be described as operating 'a rag bag of alternative systems' – it is the UK. The question surely is why? Could it be constant client dissatisfaction? Under the heading

Table 8.1 Procurement types

Country	Lump sum	Managed separate trades	Design/ build variants	Management contracting	Partnering PPP/PFI
Belgium			√√√		
France		√√√		√	√
UK	√√√	√	√√	√√	√√

√√√: popular; √: seldom used.

of civil law countries, the procurement practice of two EU states will be considered – France and Belgium.

France

France is a civil law country. For those who come from a different law background it is easy to miss the fundamental difference in approach between the Law of Contract in civil law systems and common law system. In essence most common law systems do not have a Contracts Act and it is open to the parties to a contract to stipulate a partition of risks and liabilities which are wholly contrary to the provisions of the common law. Provided that such provisions are not illegal then they will be, in general, valid contractual provisions. In civil law countries the law of contract has been codified and as a generality that codified statement has precedence over the stated intention of the parties. Common law lawyers draft contracts to make express distribution as between the parties to a contract of the risks, obligations and liabilities foreseen to arise in the performance of the contract, for example the JCT (98) Standard Form of Contract. This approach makes no sense to the civil lawyer, for whom the legislation is the law and the risks, obligations and liabilities fall to be determined by the provisions of the legislation and not by reference to the intensions of the parties. One result of this is that contract forms in the civil law countries tend to be short and centred on the administration of a contract rather than containing the meet of the agreement between the parties. This does not mean that the parties are undertaking reduced obligations, but rather that the whole range of legislative provisions falls to be applied to the performance of the contract.

France has one of the world's largest and most efficient construction industries and so-called 'machines' like the gigantic Bouygues have a large share of the domestic market. In fact, a brief look at the structure of the Bouygues Group seems to personify the dynamism and flair of the best of French construction. Founded in 1952 and still family-owned Bouygues has grown in 50 years of trading, into a diversified industrial group with interests in telecom, the media and public utilities management as well as construction and civil engineering, allowing it to better ride out any turbulence in specific markets, or the economy in general. Along the way came such prestige projects such as Parc des Princes, Charles de Gaulle Airport, L'Arche de La Défence and the French National Library as well as major works throughout the world. The French National Library was conceived and built around the same time as the British Library at King's Cross, London, but unlike the UK project the French library was completed on time and to budget.

As Professor Graham Winch's study, *Edkins, A.J. and Winch, G.M. (1999) Project Performance in Britain and France: The Case of Euroscan, Bartlett Research Paper 7*, demonstrated, the French construction industry appears to be able to procure buildings more cheaply than in the UK. What is more, given that the buildings at the centre of Winch's study were identical in terms of size and performance, it also follows that, in the Euroscan case the client received greater value for money from the French project than from the British.

The raw data contained in Tables 8.2 and 8.3 does of course need to be treated with caution. For example, the cost of employing labour in France appears to be very high compared to the UK, due in part to the high cost of social security payments and the onerous conditions of employment that operate there. However, these additional costs have to be incorporated and reflected in building prices generally in France and therefore

Table 8.2 Comparative building costs

Country	High rise apartments (€/m²)		Shopping centres (€/m²)		Air-conditioned offices (€/m²)	
	Low	High	Low	High	Low	High
France	740	970	660	1000	1280	1800
UK	1250	2000	1100	1650	1750	2600

Source: Construction Statistics 2002 dti/Gardiner and Theobald.

Table 8.3 Building labour rates

Country	Basic rate (€/hour)		All-in rate (€/hour)	
	Unskilled	Skilled	Unskilled	Skilled
France	7.08	8.62	13.94	19.58
UK	6.79	7.78	8.94	11.97

Source: Construction Statistics 2002 dti/Gardiner and Theobald.

one would expect to see evidence of higher turn-out costs! The cheapness of air-conditioned offices in France can be explained by French clients and end users being less attracted to a high specification air-conditioned environment than in the UK, put simply – they sweat and open the window. Even with these factors, generally the prices paid by UK construction clients is one of the highest in Europe, outside Ireland, where during the past 5 years, inflation, both retail price index (RPI) and tender price inflation has been in double digits.

	France	*UK*	*Ireland*
Building tender inflation	1.6%	3.5%	6.8% (2001)
Retail price inflation	1.7%	2.6%	15.70% (2000)

What therefore are the principal differences between approaches to procurement in France and is there anything that could be incorporated into the UK system?

1. Types of contract

There are two basic forms of contract in France.

- *Private sector*: The Standard Form of Building Contract in the private sector is based on the Cahier des Clauses Administratives Generales (CCAG) and is published by a French body equivalent to the British Standards Institute. The CCAG is a general document which it is inadvisable to use on its own and needs to be supported with a wide range of project specific documents specifically prepared for the project. This CCAG editing process is generally carried out by a lawyer. Sometimes there may be as many as fifteen contract documents that are arranged in descending order of importance. In the case of dispute the documents are consulted in descending order of importance.
- *Public sector*: In the case of major infrastructure works the Marche de Travaux Public is used which may be, but is not often modified, for private construction use. It has been specifically drafted however for large scale civil engineering projects in which the French have so much aplomb.

2. Insurance provisions

France has regulated the statutory position *vis-à-vis* liability by making provisions in the Code Civil (Article 1792).

The French system of compulsory construction insurance has a major influence on procurement. In 1978 the so-called Spinetta Law brought up to date articles contained in the Code Civil relating to construction liability and insurance as follows. The system has two parts:

- Dommages-ouvrages (DO) or works insurance taken out by the building owner. This policy guarantees immediate repair and making good to any latent defects that become apparent during the first 10 years. In practice the majority of claims are made during the first 7 years. The cost of this insurance is approximately 0.7–1.00% excluding tax, of the construction costs.
- Responsabilité decennale des Constructeurs (RC) – insurance held by constructors to cover claims made on them by the insurance companies in the event of latent defects. Constructors are defined as; all the actors having participated in the construction directly and indirectly by having a contract with the client or having a predominant role in the construction. It includes architects, quantity surveyors, technicians, selling agents, etc. The cost of this insurance is approximately 1.5–2.5%, excluding tax of the construction cost.

The insurance system is policed by a technical auditor or contrôle technique whose principal function is to minimize the exposure to risk for the insurance companies.

Within France the Spinetta system is sometimes criticized for:

- its complexity, two insurance policies being required,
- its inflexibility and lack of adaptability to cover specific risks,
- the cost – the total amount of premiums paid annually for construction insurance in France in 2000 was in excess of €1000 million of which approximately €800 million was from Responsabilité decennale des Constructeurs and €250 million was Dommages-ouvrages.

In addition other consequences of the system are seen as:

- The mitigation of responsibility of the client, assured as they are by the 10-year guarantee. Equally, clients could be encouraged to use cheaper approaches to construction,

particularly in the public sector, where bottom line costs are still important.
- A loss of 'professional conscience' by some contractors knowing that in the long run the insurance companies will pay.

Despite the perceived disadvantages of the French insurance system the advantages can be said to be:

- Unlike the UK system, French contractors have developed an approach to procurement that allows them to take concept designs and to re-engineer them, to produce a value for money solution. This re-engineered project then forms the basis of the contractor's bid. In theory, despite the criticism outlined above this process should ensure maximum buildability into the project.
- The French contractor is able to adopt this approach because of the Spinetta insurance system that covers the client in case of subsequent defects.

A primary difficulty with fixed price period guarantees is that no one can accurately predict what will happen within the world's economy over the next 7–10 years. As the insurance and reinsurance industries are very closely aligned to the world economic cycles any mistakes in such forecasting can have dire consequences, as outlined in Chapter 2. An additional problem such insurance faces, is the changing nature of the construction industry. If we look to the past, industry has changed considerably. How will such change, of which we are currently unaware, impact on the insurance? For example, one recent major hicc-up in the French insurance system came between 1990 and 1994 when substantial decreases in construction activity resulted in a consequential rapid decrease in the amount of premiums coming to insurance companies, while the amounts paid out for defects, clients remained relatively constant.

More than any other factor, it is undoubtedly the insurance provisions that ensure contractors are major players in the procurement of added value projects in France.

The technical audit or contrôle technique

The technical audit is carried out on behalf of the insurance company usually by an engineer working for a large bureau de contrôle such as SOCOTEC or Bureau Veritas. These bodies are private sector organizations but heavily regulated by the state and have the responsibility for checking drawings and other technical details crucial to the integrity of the building. There are those who argue, based on the 'too many cooks' theory, that this extra pair of eyes, although ensuring structural integrity, fails to raise the overall level of quality of the finished product in France.

Construction professionals

Engineers are the lead professional in France. Their education, position and status can be traced to various European military campaigns fought by Napoleon in the 18th and 19th centuries.

Architects in France have traditionally been viewed from the point of view of a fine arts practitioner rather than a specialist professional or technician. However, in 1968 the normal educational route into the profession was changed with 22 new schools being set up, and since 1977 anyone wishing to practise architecture in France had to be registered with the Society of Architects. Currently approximately 80% of the membership of this society are in private practice which tend to be small, that is less than five architects. As previously stated architects' involvement in the construction process tends to be limited to sketch designs and some basic detailing, they do not normally produce working drawings. Compared to the UK the French architect is less broadly based, both in their education and functions with little emphasis on management. The bidding system often allows for alternative designs to be submitted by the contractors. Traditionally French tender documents rely heavily on the written description, with drawings being of less significance. The drawings that accompany the tender usually being the planning drawings.

The technical design offices (Bureaux d'Etudes Techniques – BET) These organizations can carry out a function similar to that of a consulting engineer, they are often linked with, or owned by, the large contracting companies, most, but not all, are commercially oriented. They can be multi-disciplinary or they can be highly specialized. They have a strong technical expertise and are as fully conversant with the respective codes and regulations as their UK equivalent engineers. They generally provide the hardcore working details and production information.

The client, architect or contractor can employ a BET. One of the principal objectives of a BET is to maximize economy in the design from the given information. Not unnaturally, a BET employed by a contractor will tend to pursue the cheapest solution for the contractor that may not necessarily be the cheapest and the most appropriate for the client. It is common for the contractor to execute the production drawings for the project due in part to his design responsibility. The degree to which design responsibility is apportioned between the consultants and the contractor will depend upon the mission of the consultants, which in turn will be influenced by the system of contracting. There are 12 mission categories that cross-reference scope of work sub-sections to suit the client's particular needs. The contractor has design responsibility.

Quantity surveyors are not widely used in France, outside large cities such as Paris or Lyon. For the past 30 years or so several large UK quantity surveying practices have tried to break into the French market, but with little success.

Approaches to procurement and contracting

Typically therefore the French procurement process is as follows, based on a hypothetical $5000\,\text{m}^2$ office block for owner's occupation.

1. The client has decided to follow the general contractor approach and appoints an architect to prepare concept drawings and outline specification. The architect or engineer normally prepares the tender documentation with the assistance of a lawyer and the French equivalent of a quantity surveyor.
2. A short list of contractors is drawn up and the tender documents are dispatched. The instructions to the tenderers include the provision for the contractors to submit alternative solutions or method statements for all or part of the project. The idea is to allow the contractor to re-engineer the process in order to obtain maximum buildability, a process impossible under the current UK liability and professional indemnity insurance provisions. It may be that one contractor may decide to submit a price based on pre-cast concrete while others may choose *in situ* concrete or steel. Each bidder will prepare a bill of quantities, but on a different basis

both in terms of items and pricing. The quantities contained in these bills are not binding but the rates usually are. Another approach favoured in certain regions of France is for a quantity surveyor or measurer to prepare detailed bills of quantities that include full details of the manner in which the quantities have been prepared, expressed to three decimal places. These detailed bills of quantities are then sent to selected contractors, together with drawings and they are asked to check the quantities for accuracy and then add prices to form their bid. If the contractor fails to challenge the accuracy of the quantities at this time then the contractors must stand by them for the duration of the contract.

3. The bidders contact a BET, or in the case of a large contractor they use their own in-house BET and the bids and proposals for the project are produced.
4. The client receives the bids, in practice it may be that each bidder has prepared a submission of a different approach to the project.
5. The preferred bidder is chosen.
6. The client and the selected bidder often enter extensive post-tender negotiations to reduce items such as the completion date, contract sum, etc.
7. In many cases, even with large companies like Bouygues, an advance of up to 15% of the contract price is paid to the contractor, set against a bank guarantee. A substantial advance payment may result in a reduced contract sum.
8. A period of preplanning is allowed by the contract, usually 30 days, in order that all necessary information is available when work commences on site on day one. The contract would normally contain a list of the documents that the contractor will have to prepare during this period.

Contractual arrangements

Contractors at all levels for public works are required to have qualification status and are rated on a national register according to their experience, range of work, number of staff and turnover, the French equivalent of KPIs. In the public sector the client must follow a mandatory procurement procedure whereas the private sector is markedly less prescriptive. In general, three methods of contracting may be identified, the third being a variant of the first.

Separate trades The traditional way of procurement in France and most widely used. It consists of awarding separate contracts or work packages (lots) to each specific trade or groups of trades. A work package may involve the services of several specialists that work together and the number of packages varies from job to job. Using this approach each trade contractor has a direct contractual relationship with the client, but no legal ties with other specialists. A project manager is usually appointed to co-ordinate the packages. The policy of dividing the work into packages not only satisfies the technical aspects of the projects, it also enables smaller companies to compete for work on larger projects – one of the reasons why the smaller trade firms thrive in France. Advantages include hands-on control by the client and high quality finished product. In addition the practice of inviting tenders for earlier parts of the project prior to that of later parts and usually before the overall design has been completed, provides a degree of fast tracking and parallel working. Disadvantages arise if one firm delays site operations, as none of the other firms has an obligation to rectify the situation and this can cause disruption to ripple through the whole project.

General or main contracting Becoming more popular as a one-stop shop for construction projects. The advantages are less risk for the client due to delays by the works contractor. Disadvantages include lack of control by client or even design team over the contractor and the tendency to make the job fit the price with subsequent reduction in quality of the finished product.

Grouping of contractors Two contractors group together to pool their expertise. They may share risk in a variety of ways.

The structure of the French construction industry is similar to the UK, that is to say a large number of small firms and a few large ones. However, due to the separate trades system of contracting the role of the small firm is perhaps more important than the UK. The amount of French sub-contracting is small compared to the UK and there are no nominated sub-contractors, as design and construct is the predominant form of procurement.

The open competition system is still widely used in the procurement of French public sector assets. Each week *Le Moniteur*, the most widely read construction industry journal in France, has large numbers for calls for submissions for a wide variety of projects advertised on a region/departmente basis. The competitions vary and can be from professional services to carrying out the works. Although on the face of it, a very good and transparent way to enter for work in France, the information that has to be submitted is often highly prescriptive and the choice of winning candidate can be very political with the mayor and local officials often exerting influence in the selection.

Finally, the French can be justly proud of the system of public sponsored research based in the many specialist higher education establishments throughout France.

Summary

The range of procurement strategies are more limited than in the UK being in the main, competitive, selective and negotiated – see Table 8.1. However, the French procurement process is very different from the UK and appears to be less confrontational and gives the client more security, although whether the French client receives better value for money is difficult to say.

- Contracts start from a position of a more positive approach and there is less of a 'them and us' situation.
- Tender periods are shorter and in reality the system mitigates against the lowest bid being accepted *per se*, as the tenders often relate to different approaches to the project. Tender appraisal is very much a key function in deciding value for money and often contracts are placed following extensive post-tender negotiations.
- Design and construction are well integrated.

Belgium

Like France, Belgium is a civil law country but has developed in a slightly different way. Lessons from history show us that the British and Belgian construction industries at different points in time, unlike any other national construction industries in Europe, set off in two unique and quite distinct directions. Whereas the UK developed an environment that regulates construction activities, Belgium decided instead to regulate the actors. In the UK the so-called 1828 Committee of

Enquiry set up by Parliament in that year decided to switch all public works from separate trades contracting, described in more detail in the section on France, to contracting in gross. The so-called Great Revolution in Contracting set the seal for the UK construction industry to set off down a path and into a working environment that still today distinguishes it in a number of important respects from all others.

In Belgium in 1934, following a series of accidents, failures and bankruptcies, the government, as governments do, threatened to intervene with regulations, to protect the interests of owners and general public. However, in December of the same year Professor Gustave Magnel of Ghent and Eugene Francois founded a control body known as SECO. This non-profit making organization stems from the three essential professions in the field of construction, namely: architects, consulting engineers and contractors and is owned by the industry.

So how is the Belgian construction industry regulated? The simple answer is – and for those anglophiles who detest regulation this may come as a surprise – hardly at all. Firstly, Belgium is the only country in Europe to have no technical building regulations. On the other hand, Belgian law prescribes that no building, new or refurbished, may be erected without the intervention of a qualified architect. So how does all this fit into the procurement process? Like France, project insurance is available from a number of insurance companies at a cost of little more than 1% of total construction costs. Unlike France however, it is voluntary and in practice only about 20% in value of all contracts use insurance, the most large and/or complex projects. For contracts without project insurance rely on the design team's professional indemnity insurance and a system of joint and several liability. If SECO is used then it is normally employed by the client and linked to a system of project insurance. Under these policies all parties including the contractor are covered against design faults and latent defects. As in France a technical audit is carried out throughout the design and construction process.

What Belgium has attempted to do is to regulate participants in the procurement and construction process this includes the use of only:

- Approved contractors
- Degree qualifications for all professional advisors.

Table 8.4 Belgium and UK compared

UK	Belgium
Fundamental differences	
• Risk shedding	• Risk sharing
• Cost control	• Cost reduction
• World-class designers	• World-class innovators
• Over design by consultants	• Lean design by consultants
• Over specification	• Performance specification
• Bespoke solutions	• Flexible but standard solutions
• Problem solving	• Problem elimination and avoidance
• Prone to conflict	• Conflicts inherently avoided
• Labour intensive	• Capital intensive
• Low level of capital formation	• High level of capital formation
• 'Cowboy syndrome' is prevalent	• Qualification and registration reduces 'cowboy' tendencies
• Time spent negotiating contract conditions and warranties	• Project insurance risk cover facilitates rapid conclusion of negotiations
Organizational aspects	
• Low productivity	• High productivity (+30%)
• Strengths in design and management	• Strengths in site organization and management
• Weakness in site operations	• Strengths in site operations
• Premium on flexibility	• Emphasis on high productivity
• Reliance on sub-contract and/or labour only sub-contractors	• Reliance on 'in-house' teams of highly trained individuals
• Detailed bespoke designs	• Performance specifications and standardized components
• Many consultants' drawings; few shop drawings	• Few consultants' drawings; more contractor's execution drawings
• Longer up-front design periods	• Shorter lead-in times
• Little requirement for in-house design capability	• In-house design capability essential to innovation and winning work
• Lowest price often based on bills of quantities	• Emphasis on alternative technical solutions to win work
• Many site supervisors	• Fewer site supervisors
• Weaknesses in management skills	• Management skills perceived as essential to success
Social aspects	
• Low wages fail to attract quality employees	• Good wages attract better quality employees
• Cheap labour = higher out turn costs	• Expensive labour = use sparingly = lower out turn costs
• Low rate of permanent employment	• High rate of permanent employment
• Short-term fragmented teamwork = discontinuity	• Long-term cohesive teamwork = continuity from site to site
• Inadequate training lowers productivity	• Emphasis on training

Source: John Goodall, FIEC, Brussels.

With these controls in place the voluntary addition of latent defects insurance is seen as a positive selling or letting argument for the client. How has this development manifested itself in construction procurement in Belgium? Table 8.4 is a personal view compiled by John Goodall of the FIEC that is based in Brussels.

How therefore could the UK construction industry benefit from the French and Belgian construction industries and how in turn could this lead to added value in procurement? It is of course unrealistic to imagine that it would be feasible or even desirable to transform the environment in which the UK construction industry operates into a French or Belgian one; however, the following items are noted here for consideration.

The introduction of project insurance. The counterpart of such an initiative is that one or more inspection bodies, possibly along similar lines to SECO would need to be put in place, essentially to protect the interests of the insurers. Standard forms of contracts would need to be amended to allow scope for the contractor to modify the technical design prior to construction, with consultant architects and engineers waiving their right to interfere. Table 8.5 summarizes other EU countries' approaches to regulation and insurance provision.

Table 8.5 Contracts in Europe

Countries under Napoleonic influence	
• Grand Duchy of Luxemburg	Similar to Belgium
• Spain	10 year insurance on housing similar to France
• Italy	10 year insurance Technical control on design only Limited success
Other European countries	
• Germany	Contractual reduction of liability (2/5 years) Few insurances Highly developed standards Prufingenieure – individuals approved by Lander
Northern European Countries including UK	Regulations Professional indemnity insurance Public controls – Building inspectors

Attention will now be turned to a country with a federal system, a system that presents problems particularly when it comes to regulation and implementation.

Countries with a federal system

Australia

Australia has a lot in common with Europe, in so far as Australia has nine different parliaments including the Federal (Commonwealth) Government and eight State and Territory governments, therefore from a legal jurisdictional point of view, in certain areas it could be considered to be nine separate countries. Australia also has a lot in common with the UK when it comes to the construction sector, for according to the Australian Procurement and Construction Council (APCC), in 1997 the profile of the Australian construction industry was as follows.

Out of 230,000 firms employing some 750,000 people:

- 65% of all enterprises in the industry employ less than two people;
- 88% of all enterprises have a turnover of less than $500,000 (£190,000);
- Less than 1% of enterprises employ more than 50 people;
- Only 1.3% of enterprises have a turnover of $20 million (£7.5 million) a year or more.

While this diversity makes the industry highly competitive, the less satisfactory characteristics include:

- Adversarial culture
- Under-capitalization
- Low margins, with little or no investment in research and development of new processes or use of new techniques
- Short-term focus, relationships and planning
- Fragmented approach, second only to agriculture.

Further the APCC concluded that the Australian industries' construction processes are largely determined by the industries' sub-contracting structure. This has particular implications. Based on the profile of these enterprises, namely 77,000 establishments with zero to four employees, predominantly family business and generally taking the form of a commercial venture performing a particular industry trade or task skill;

- Up to 20 specialist skilled sub-contractors may be employed on a residential housing project.
- Up to 200 specialist skilled sub-contractors may be employed on a major project.

This results in a system where work is organized into small, almost isolated packages. The outcome is a fragmented approach both in terms of design, where separate small design consultants are used project by project and in terms of construction where multiple levels of small specialist sub-contractors and suppliers are used. This fragmentation, together with the divisions between design and construction, limits opportunity for efficiency gains and encourages the pursuit of singular interests.

Like the UK, Australia also has a poor record of investment in research and development in construction. However, in 2001 the Cooperative Research Centre was initiated, charged with the responsibility of *'Leading the Australian Property and Construction Industry in collaboration and innovation.'* The centre will have an initial budget of $64 million over 7 years.

Liability and insurance reforms in Australia

In Chapter 2 the merits of introducing project or latent defects insurance was discussed in a UK procurement context. The situation in Australia pre-1993 was much as the UK today – a complex mismatch of insurances with no compulsory guarantee or extended warranties for the client. In the early 1990s the state of Victoria decided to try to improve the situation and in 1993 the Victorian Building Act came into force. In the years that followed many other states adopted similar legislation. In essence the Victoria approach is very similar to the French system.

- Compulsory insurance cover for a fixed period after the handover of the completed building.

- Automatic run-off cover as part of the insurance, enabling a client to be covered against a defect, long after the person or organization responsible has gone out of business.
- Cover is mandatory for:
  - Professional consultants
  - Commercial/industrial builders
  - Domestic tradesmen.

Premiums are calculated on the basis of a risk assessment. The result of this initiative is considered to have been a success and has resulted in a system where construction clients have more confidence in the procurement process.

PPP/PFI in Australia

Perceptions of PPPs in Australia

The world market for public private partnership (PPP) projects has been projected to be worth US $32 billion by 2007 according to John Spoehr of the University of Adelaide. Earlier, in Chapter 4, recent developments in the adoption of PPP/private finance initiatives (PFIs) by the UK government for the procurement of public sector assets was discussed. However, the popularity of this approach to procurement has spread around the world quickly, and in Australia there is increasing pressure on public sector agencies to adopt the PPP/PFI approach.

The major drivers for the adoption of PPPs in Australia are thought to be:

- PPPs are attractive to state governments strait jacketed by their commitments to no new taxes and lowering public debt.
- 'Off balance sheet' PPP deals create the impression that public sector debt levels are declining.
- The tantalizing prospect for PPP advisors, that is consultants, lawyers, etc., of 15% returns on PPP projects, compared to the returns available in the more mature markets of say the UK of between 10 and 12%.

The formation of PPPs in Australia occurs largely on a case by case basis and is officially driven by economic considerations, which can be quite different in nature, depending on the circumstances. In the recent past, since 2000, PPP/PFI deals have started to gain popularity with the various Australian governments. Unlike the UK there is no over-arching policy in this respect, but there are several legislative constraints which apply. The choice of the terminology 'Private Financing' in Federal Government literature is interesting; however, as it places emphasis, quite correctly some may say, on the source of the finance for these type of schemes. However, it seems to fly in the face of the UK experience that has after years of adverse criticism, that the private finance of public services was a mere gravy train for the private sector, sought to soften the image of the initiative with the title PPP.

The Australian Federal Government set out the core principles of private financing as:

- value for money as tested by the use of a risk adjusted public sector comparator,
- transparency through the disclosure of privately financed deals,
- accountability through established best practice procedures

and therefore in many respects is very similar to the UK model.

In spite of this central guidance, public infrastructure responsibilities in Australia are fragmented as a result of Australia's development from a number of colonies to a federal Commonwealth. After federation in 1901 responsibility for some infrastructure matters passed to the Commonwealth or Federal Government, while the majority remained with the states/territories. Some matters are devolved to the local level as follows:

- Federal responsibilities
    - Telecommunications – now deregulated
    - Aviation
    - Shipping
    - Involvement as facilitator, but not owner or participator, in main road networks, interstate electricity, gas pipelines and rail networks.
- State/Territorial responsibilities
    - Rail, roads
    - Supply of water, electricity and gas

- Development and maintenance and operation of inter-state roads and rail links.
- Federal/State shared responsibilities
  - Funding for major interstate projects such as highways.
  - In addition to the above a fourth level of administration applies.
- Local government responsibilities
  - The local provision and maintenance of local roads.

The growth of PPP in Australia

Most state governments now have PPP policies and programmes in operation; however given the above legislative structure the application of PPP/PFI in Australia varies from state to state and state governments do not need to obtain federal govern-ment permission to proceed with a PPP project. The application of PPPs in the following Southern Australia states will be briefly discussed:

- New South Wales (NSW)
- Victoria
- South Australia

South Australia   Most analysts agree that South Australia's public infrastructure is ageing and in need of modernization after more than two decades of under-investment. The need for substantial new infrastructure investment is pressing, but resources available through state budget are limited and bor-rowing is unfashionable. South Australia has therefore decided that the future prosperity of the region will depend on a sub-stantial increase in public private infrastructure investment over the next 5–10 years. In September 2002 Partnerships SA was launched, described as a procurement programme for the private and public sectors that seeks to promote private sector participation in the delivery of government services to the community where there are sound reasons to support this approach. The procurement process laid down by the South Australia government is illustrated in Figure 8.1.

Of all of the States courting PPP, South Australia's approach and programme appears to be the most organized. However, despite the criticisms levied in the UK at the Public Sector Comparator, see Chapter 4, South Australia, in common with

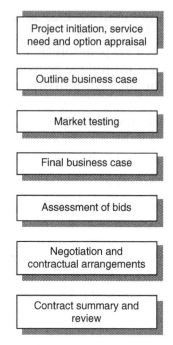

Figure 8.1 PPP/PFI in Australia.

other states has chosen to adopt this model for demonstrating value for money.

New South Wales   Publicly financed project guidance has been available in New South Wales since 1989; however in 2001 the Private Finance Policy was revised and re-issued. Under the new policy Privately Financed Projects (PFPs) are required to demonstrate a net benefit to the community. The revised PFP policy seeks to engage the private sector in a range of activities including social infrastructure. A key criterion in the NSW policy is demonstrated value for money involving:

- Improved risk management
- Ownership and whole life costing
- Innovation
- Asset utilization
- Whole of government outcomes.

Once more the ubiquitous public sector comparator is the value for money model.

To date the experience of PPPs in NSW has not been good. For example, the Sydney airport rail link, opened in May 2000 and operated by a consortium including the French giant Bouygues ran into financial difficulties when estimates of passenger numbers proved to be widely over-optimistic, forcing the consortium into receivership. On a more optimistic note, at least 80 financial institutions and construction firms expressed interest in a plan to build and operate nine new schools in Sydney's northwestern fringe, in areas experiencing rapid population growth. It is thought by some commentators that using the PPP approach in this scheme will open up an entirely new approach to the establishment and maintenance of social infrastructure in NSW.

Victoria Partnerships Victoria was launched in June 2000. In essence the rhetoric in the state government's literature describing the scope and application of PPP are striking similar to the information available from the UK Treasury. However, two interesting inclusions are the sections on; protecting the public interest and an assurance that money and resources will not be wasted on schemes that are not 'PPPable.' Indeed concerns have been raised about the viability of some of the markets in terms of their size and whether some of them are subcritical. If this should turn out to be the case then there is speculation that future price rises will exceed pre-PPP prices within the next few years. The flagship of Victoria's PPP schemes is Spencer Street Station due for completion in 2005 in time for the 2006 Commonwealth Games. The project involves the construction of a $350 million station together with its operation of a 30-year contract period.

Procurement lessons from Australia

A more guarded approach to PPP?

Opinion surveys always show that between 60 and 70% of Australians are opposed to the privatization of public services; an opposition that stretches across all demographics and voting intentions. It is also clear that governments at all levels in Australia from Federal to local level are determined to press on with the use of PPPs as the preferred method of procurement.

In addition to the cases referred to above there have also been some very high profile healthcare PFI failures in Australia namely; the Joondalup Health Campus, Western Australia and the La Trobe Hospital in Victoria, where, 4 years after signing the contract, the hospital was handed back to the Victorian Government in October 2000 with a reported annual loss of $6.2 million. In December 2000 Senator Rosemary Crowley chaired an extensive enquiry by the Senate Community Affairs References Committee into the funding of the Australian Health Service and came to the conclusion in the final report Healing Our Hospitals: a report into hospital funding that *'No further PFI schemes should occur until a detailed review has been undertaken and benefits for patients have been demonstrated.'* Although this recommendation is unlikely to be implemented it would appear that there are no moves afoot to turn PPP/PFI into the only procurement game in town as is the case in the UK. Australia has also demonstrated that a system based on compulsory insurance can be introduced into a system where it did not exist before and can make the procurement process more certain for a whole range of clients.

Bibliography

Franks J. (1998). *Building Procurement Systems – A Client's Guide.* Longman.
Goodall J. (2001). *A personal view from Brussels.*
Spoehr J. (2002). *PPPs in South Australia.* Evatt Foundation.
Crowley R. (2000). *Healing our hospitals: a report into hospital funding.*

Web references

www.seco.be/
www.aphgov.au/

9
Procurement – the next steps

Introduction

Previous chapters have described a range of new strategies currently being introduced to procure built assets, most of which have a common denominator of non-adversarial team centred methodologies, based on trust and respect. However many within the industry would agree with Digby Jones, the Director General of the Confederation of British Industry, when in an article in *RICS Business* he accused the construction industry of *'talking the talk about change but not walking the walk.'* Chapter 1 drew attention to the recent proliferation of recent government initiatives, announced in an attempt to change the 'walk' of the UK construction industry, but for the majority of construction industry, that is to say SMEs, putting many of these initiatives into practice stays on the back burner. In addition to all the government sponsored initiatives, sources in the private sector have called for a move towards world class performance and lean production philosophies – can these seemingly ambitious targets and ideals be achieved?

Obstacles to improvement

As Zara Lamont remarked in the Foreword: *'The strongest impression that has been left with me is how markedly the language and culture of the industry is changing towards the acceptance of a fully integrated team approach in identifying clients' construction needs and satisfying them in ways which add value to their core business.'* Is her optimism justified? Generally, yes, the bigger picture does look encouraging; however, obstacles remain to widespread adoption of new procurement strategies. Not least of these is that the construction industry in very conservative; observe the howls of protest when GlaxoSmithKline introduced e-procurement a few years ago. Coupled with this is a suspicion among certain sections of the supply side of the industry that there is less than total commitment by clients to new approaches to procurement and there are a number of hidden agendas. For example, consider partnering, the backbone of many new approaches to procurement; many within the industry have questioned the true commitment to this new approach to procurement shown by some clients with the observation: 'is it true partnering or something which tries to meet current fashion' and that many organizations are merely paying lip service to it. In addition, the Civil Engineering Contractors Association (CECA), in its report Supply Chain Relationships (2002) made the point that, in addition to the apparent perceived lack of commitment by clients, in the civil engineering industry's marketplace, there have been a number of other fundamental obstacles to the development of, for example, long-term partnering. These include principally the following:

- *Fluctuations in workload*: Construction is a cyclical business, which in the private sector has been traditionally, but not exclusively, linked to the economic cycle. It is difficult to keep supply chains intact and working to full capacity during sudden decreases in workload and confidence. In addition some elements of public sector works, for example road building, has been closely linked to government policy. The period between 1990 and 2010 has seen the actual and planned investment in new roads fluctuate from £1.5 to 0.7 billion by the turn of the century rising to £1.7 billion a year in 2008/2011.
- *Client base re-organization*: Changes of ownership both in the private sector and re-organization of the client side in the public sector can interrupt not only the flow of work but also can disrupt normal working relations between contractors and their clients.
- *Changes in government policy and in legislation affecting client behaviour*: There have been changes in policy and legislation that have affected relationships between contractors and their clients in both the public and private

sectors and also between contractors and their suppliers. Privatization, sustainability considerations, and the initiatives promoted by the Byatt Report are discussed later in this chapter.

Perhaps the NHS Estates have the only approach likely to achieve change in such an intransigent and conservative industry as construction, namely to adopt a procurement policy with few options. The message from NHS Estates to industry is clear: if you want you do business with us prove your commitment to non-adversarial procurement paths, otherwise don't bother. A similar position has been taken by the Ministry of Defence. In the roll-out of the NHS Procure 21 strategy it was recognized at an early stage by the NHS Estates that change would not come from the supply side of the industry – it would need to come from the client and a widespread campaign to turn NHS Estates into a best client was instigated and is described in Chapter 5.

Of all the routine business procedures, procurement is the one that has proven to gain most, in added value terms, from the introduction of electronic formats. For since the mid-1990s the future of procurement would seem to be inextricably intertwined with information technology and systems.

After the hype – was it the reality? Reports from around the world towards the end of the 20th century predicted vast increases in the use and the efficiency of e-procurement. Firstly, a look at the impact of e-commerce on UK construction generally. In 1999 the Government's Performance and Innovation Unit identified three priorities for the UK to be 'the best place in the world to do business electronically by 2003'. These were:

- to overcome business inertia,
- to ensure that government action drives the take-up of e-business,
- to ensure better co-ordination between government and industry to gain maximum benefit from existing and proposed programmes.

The above statement would seem to imply that in 1999 at least, the UK Government was confident in predictions that the rapid adoption of e-commerce by industry and official figures and estimates at the time appeared to confirm this optimism.

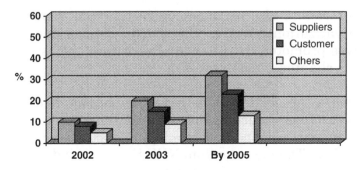

Figure 9.1 Construction companies using e-commerce (source: CPA/CIPA, 2000).

In the UK in 1999 e-business transactions, at the time of dot.com mania was £2.8 billion with the potential for growth being esti-mated to be ten-fold in the following 3 years. As for the con-struction sector forecasters predicted that by 2005, 50% of all transactions could be undertaken by e-commerce – that has since been reduced to a more cautious 22%. A survey carried out by CPA/CICA in 2001 and 2002 produced the results illus-trated in Figure 9.1.

Certainly, in the few studies that have been undertaken into the impact of e-commerce in construction, project tendering, online bidding and procurement have been identified as one of the major kinds of construction activity to emerge, that have adopted an electronic based approach.

e-Procurement

Since 1998 many claims and predictions have been made for the return on investment rates achievable through e-procurement projects. By mid-2000 these figures had reached 'between 200 and 400% according to *Buy IT*. A year later a random survey of procurement managers using a consistent model indicated a potential average return on investment of 8% from implemen-tation of e-procurement best practice across a range of organi-zations. Today, as the first generation of e-procurement projects should be starting to deliver measurable savings, the picture as to the real extent and value of these savings is still unclear. One explanation for this is that while project progress has been measured in terms of milestone achievement, relatively few organizations are accurately monitoring the real benefit achieved as the projects progress.

The CPA/CICA study concluded that issue of processing of tenders occurs mainly between procurement teams and contractors and also between contractors and sub-contractors. Tools are already available that allow the analysis and processing of tender documentation. The issues in this area seem to centre on greater uptake rather than radical innovation. Preferences for paper based tendering do not necessarily rule out the use of e-tendering by e-systems which must be proven to be as robust and secure as the traditional paper based approaches. It would seem in the absence of significant examples of large projects using e-tendering that the widespread use of this application is still far off. However, there are other aspects of the procurement process that can and do utilize e-commerce. As previously discussed there is an increasing tendency to prequalify organizations for inclusion in frameworks, etc. much of the prequalification process can be carried out electronically, saving time and effort.

There is increasing evidence that the direct economic benefit of project tendering may be relatively small. A report published by the Building Centre Trust (2000) seems to suggest that most of the savings in e-tendering come from the reductions in printing/distribution costs; there is no clear evidence yet that e-tenders are easier to complete, submit or analyse than the traditional paper based approach. In Chapter 7 Rory Lamont uses a case study to illustrate BP's experience in reverse e-auctioning to purchase specialized materials/components. One of the principal criticisms of e-auctions is that this approach is harmful to the buyer–supplier relationship. The general opinion of this approach to procurement is that if e-auctions are used to cut margins significantly they will have a dramatic effect on supply chain relationships. If the product or service cost is being cut, then surely the natural consequence would be lower quality, poor delivery performance, etc. In an age when modern supply chain management is trying to create a win/win model, to impose an openly adversarial tool seems to go against the grain. A recent survey (April 2003) carried out by the Chartered Institute of Purchasing and Supply (CIPS) in collaboration with the University of the West of England and supported by Oracle and BT, attempted to confirm the popular perception of e-auctions (*Douglas A*). Unsurprisingly, the results of the poll confirmed that e-auctions resulted in lower prices (94% of respondents), however surprisingly, many organizations that used e-auctions for the first time reported dramatic improvement in their suppliers' performance for the five key performance factors: account management, flexibility, quality of product or service, delivery, and reliability and dependability. Other findings include an increase of 22% in supplier flexibility and a 20% increase in quality of product or service or both. Therefore, on the face of this survey, the findings of the report contradict the perceived wisdom that e-auctions were detrimental to supply chain relationships. As was emphasized in the BP case study, for the e-auction to be a success a lot of preparatory work needs to be carried out in terms of which items to include, selection and training of bidders, etc. In conclusion the CIPS survey sets out some do's and don'ts of e-auctioning. Do's include; selecting contracts with a value in excess of £50,000, and include between three and five suppliers in the process. Don'ts include supply chain managers bidding to encourage a lower price or charging suppliers to participate! (see Table 9.1).

The types of benefits that are thought to arise from the introduction of e-procurement are as follows (Figure 9.2):

- Hard, that is directly measurable ones such; as price savings and process cost reduction, (head count).
- Soft, indirectly measurable benefits, for example, individual time freed up through more efficient processes.
- Finally, there are a number of intangible benefits such as cultural change.

On a more positive theme a report by the *Buy IT* Best Practice Network – Online Auctions in 2001 concluded that e-procurement and in particular online bidding can deliver benefits with a well constructed implementation plan and careful management of the process. Online bidding is seen by *Buy IT* as an efficient mechanism for purchasers and suppliers to open up competition and find the competitive market price over a sustained period of time. Whether such savings will be sustained over a long period will be down to how the major companies then deploy process improvements and supplier development tools to drive cost out of the supply chain for good. Early adopters are now using the process over a vast range of products and services and concluded

Table 9.1 Do's and don'ts of online auction

Strategy	Preparation	Event	Follow up

BUYER

Strategy	Preparation	Event	Follow up
Is this a core competence for my business? How does this fit with other procurement activities? What percentage of spend is e-auctioning? What impact will this have on key supplier relationships? Do not plan to reverse auction everything.	Decide the number of suppliers to invite Ensure sufficient market competition Train suppliers Set clear rules Do not plan the event at the wrong time – e.g. a public holiday Agree evaluation criteria Set opening price	Ensure proxy bidding process in place Monitor supplier bidding Monitor technology reliability Monitor bidding tactics Do not get carried away with the hype – is the lowest price the best?	Finalize sourcing decision Give feedback to suppliers Capture knowledge gained

SUPPLIER

Strategy	Preparation	Event	Follow up
What do I want to achieve? How much will this impact on my supply chain? How best to respond? How will this impact on my customer relationship?	Respond promptly to all buyer requests	Have your first bid read before the start of the event	Provide cost breakdown if requested

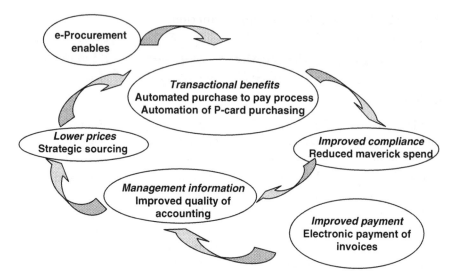

Figure 9.2 e-Procurement drivers.

that online auctions can be a useful first step for organizations keen to embark on e-procurement.

Benefits to both buyers and suppliers include:

- Transparency, highlighting the integrity of the buyer's sourcing decisions.
- Reduced lead times.
- Availability of the 'market price'.
- Encouragement to suppliers to bid regardless of location.
- Better standards of specification.
- Best total cost, rather than cheapest – the technology now enables a range of factors to be considered and weighted, for example quality, delivery and specification.

An interesting view of the coming together of e-commerce and emerging business practice, such as partnering, is given by Dan Blacharski in his article: 'No Speed Limits: Hyperpartnering in the Internet Age.' Blacharski concludes also that despite predictions of rapid growth e-commerce, is still in the 'early adopter' stage. In order to become involved in business-to-business (B2B) e-commerce, such as e-auctions, all the partners of the supply chain must be on the same page technologically and logistically too. From the supplier's side having a long-term contract is very attractive; however, the prediction by Blacharski is for less and less of these with a change over to short-term dynamic supply deals. Instead of the 'old fashioned' model of long-term contracts and linear supply chains, the linear supply chain will be replaced by a completely new market structure, called by Blacharski the 'e-business network', in which partners can switch allegiances without cost. The e-business network is really a group of inter-dependent players that cooperate with one another in real time over the Internet. This model, known as hyperpartnering, keeps costs down in several different ways by:

- allowing companies to take advantage of the spot market and day-to-day fluctuations in price,
- implementing technology to automate the workflow of the supply chain,
- helping companies find absolutely the best deal available at any given point in time,
- avoiding the need to carry high levels of stock,
- buying goods and materials on a just-in-time, as-needed basis.

In this model, businesses are not strictly competitors, but rather a combination of cooperation and competition. The Japanese business culture of keiretsu, described in Chapter 1 operates in a similar manner. Once in the keiretsu you have a solid group of associates and customers who are obligated to buy from other members, even if a company outside of your keiretsu has a better price. The model in the West is expected to develop in a slightly different way where the affiliation is a lot looser and there is probably more competition than cooperation, but the connection is still there.

The Byatt Report and local government procurement

In the summer of 2000 the Local Government Minister and the Chairman of the Local Government Association asked Sir Ian Byatt to chair an independent taskforce to review procurement processes in local government in England and Wales. The terms of reference were, to review the state of procurement and com-missioning skills and practice in local government in the light of the requirements of Part 1 (Best Value) of the Local Government Act 1999 and its objective of continuous improvement in the economy, efficiency and effectiveness of local services as well as to make recommendations. This taskforce demonstrates the determination by the government to roll out new approaches to procurement in the wider public sector, in the same way that public private partnership (PPP)/private finance initiative (PFI) was extended under the PPP Programme. The issue of the Byatt Report (*Delivering Better Services for Citizens*), and the subsequent response by the Government and the Local Government Association in July 2002, focused the attention of local authorities on the subject of procurement. In particular, it highlighted the need for authorities to widen their horizons by taking a corporate view of procurement issues. This includes a shift in emphasis from initial purchase costs and short-term savings to the examination of whole life costs and establishing longer-term objectives to secure overall best value. Coupled with the Byatt Report are the various initiatives associated with the Egan *Rethinking Construction* agenda. In this respect, considerable work has been undertaken since March 2000 by the local government task force (LGTF) to promote the *Rethinking Construction* principles within local authorities. Procurement makes a critical contribution to the quality of public services and the total expenditure of a typical local authority on non-payroll spending (i.e. bought-in goods, works and services) is measured in hundreds of millions of pounds. The main thrust of the Byatt Report was that authorities should review their corporate structures and processes associated with this expenditure, and set out a clear policy on how procurement is to be managed across the authority. The report concludes:

> *Each authority needs to review their current procurement structures and processes and set out a clear policy on how procurement is to be managed across the Authority. All unitary and county councils will be expected to establish a corporate procurement function to stand alongside finance, performance management, legal and human resources.*

The Byatt Report suggested that the procurement strategy should include general procurement principles and methods as well as updated information on the procurement activities of the council. Information which could usefully be set out in a strategy included:

* Strategic aims of the authority and their relevance to procurement activities.

- An analysis of key goods and services and their costs.
- Information on how goods and services are purchased.
- Details of current contracts with renewal/replacement dates.
- Recent planned best value reviews.
- Future procurement exercises anticipated by the authority.
- The performance of key suppliers.
- The structure and performance of the procurement function.
- Skills and training needs.
- Issues of probity and good governance.

The reply to the Byatt Report from the Office of the Deputy Prime Minister emphasizes the need for a culture change with local authorities by them embracing such things as long-term agreements with suppliers, selection of contractors on the basis of quality as well as price. In addition auditors are encouraged to move away from the traditional comfort zones offered by selection systems that have the objective of securing the lowest price and instead to adopt methods such as open book accounting, a comparison of the true cost of the traditional procurement against the cost of partnering and benchmarking of performance measures.

Clearly, the implications for local authorities of the Byatt Report seem to be that they will have to adopt a much more professional approach to procurement.

Best practice procurement

There are no absolute criteria to define what best practice is in terms of procurement; however here are a few thoughts.

Establishing the client's business case

For most companies or clients, their capital assets, whether they are buildings, roads, railways or operating structures represents their biggest fixed cost. Procurement professionals must therefore develop strategies that enable companies to turn a cost into a sustainable competitive advantage. Before this can be done the question must be asked; who is the client anyway? Is it:

- the end user?
- the board?

- the shareholders?
- the procurement/finance departments?
- etc.?

or is it some or all of them?

Clients should enter the construction process with a clear understanding of their business needs and the functionality they require from the finished product. However, for many construction clients, whether occasional or experienced, finds the construction industry an almost impossible industry with which to engage, facing a bewildering array of specialists. Of the many points of view on the movement towards a new approach to construction procurement, the one common theme appears to unite all sides; there is the need for a greater client focus and to devise systems that make it easier for clients to engage. The issues at the heart of integrating the procurement of built assets with client's business needs can be summarized as follows:

- Relate to customers and markets, understanding the client's value system, including:
  - Define business strategy and competitive advantage and how built assets contribute to these?
  - Use functional or output based briefs.
  - Develop client focused standardized design elements.
  - Improve understanding of business strategy.
  - Become a stakeholder in the procurement process.
  - Appoint a dedicated sector manager within the advisor's organization.
- Managing costs collaboratively with the aid of:
  - Value management/life cycle costs.
  - Risk management.
- This may sound very patronizing but do not give construction clients what they ask for – sometimes they do not really know. Question and challenge their needs.
- Utilize and integrate information technology and information systems.
- Managing human resources including partnering and long-term relationships.
- Integrating the processes throughout the entire supply chain through:
  - Collaborative design processes.

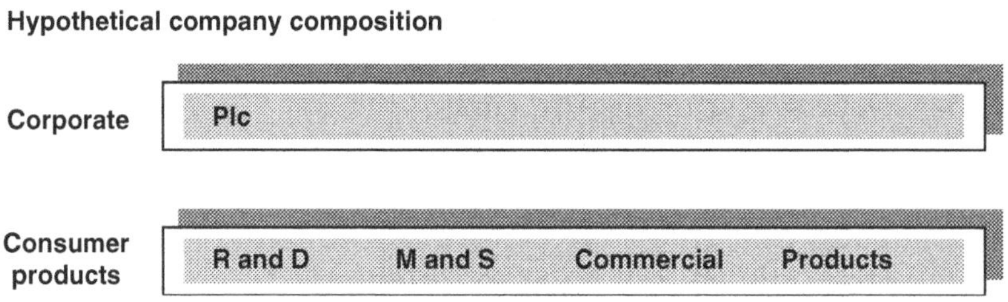

Hypothetical company composition

Figure 9.3 Procurement drivers.

Even within organizations there can be misunderstandings. For example, within a smaller organization refurbishment may be carried out and controlled by an in-house team that identify and pursue actions that are focused mostly in-house departments rather than business needs of the company as a whole. In larger organizations, the process becomes even more complex. Within a large organization there are many different parts all interrelated but all with different and distinctive drivers (Figure 9.3). To find out what clients want, it is first necessary to understand what they do and where they fit into the larger corporate picture.

Understanding the client's core business

In large organizations in order to discover what clients want it is necessary to understand where they fit into the larger picture. Within a large company there are many parts and each may have its own very distinctive and discrete drivers.

To find out what clients want, it is first necessary to understand what they do and where they fit into the big picture. For example, within a large organization there are many different parts all interrelated but all with different and distinctive drivers (see Figure 9.3).

- How does a client achieve competitive advantage?
- What is the nature of client's business?
- What are the business imperatives?
  - Identify appropriate and clearly defined project alternatives.
  - Verify that the project scope covers all that is necessary to provide the project benefits.

- Basing the procurement strategy on the entire project life cycle, giving equal rigour to operational cost and benefits as well as capital costs.
- Ensuring that there is a clearly defined project sponsor who owns all aspects of the business case.
- Ensuring that there are criteria established for measuring performance in a meaningful way.
- Align the procurement strategy to the client's business case.
- Foster openness and transparency in the supply chain.

Conclusion

By way of a conclusion to this chapter and this book here are a few observations:

- The e-commerce, e-procurement revolution has still to happen, although revolution is perhaps the wrong word as there is evidence that e-commerce is being used increasingly in day-to-day business transactions as the industry's first choice.
- There is a consensus in the industry that non-aggressive, less adversarial approaches to procurement are the way forward.
- Many basic issues still stand in the way of progress to seamless integrated supply chains, such as operate in other sectors. What is more, these issues seem very stubborn to go. For example, the habit of many prime contractors of failing to pay the supply chain members on time!
- The industry will not change without strong leadership from high profile clients, such as demonstrated by the National Health Service ProCure 21 programme.

Nevertheless pundits will probably look back at the early 21st century as the golden age of partnering and openness in the construction industry. Perhaps the first big test of the new procurement strategies will come as surely it will, when the cold winds start to blow of the first major post-Egan recession and the economy and the construction industry begin to slow down. When work becomes scarce and margins tighten, will the partnerships and alliances stand the strain? When supply chains become broken, will they be able to be welded back together to become as strong as before, or when the going gets tough will organizations go back to their old ways, muttering 'I told you so'? Let us hope not!

Bibliography

Bashford S. (2003). *Fighting Talk*. RICS Business, The Royal Institution of Chartered Surveyors. February, pp. 26–28.

Byatt I. Sir (2002). *Delivering Better Services for Citizens*. The Local Government Association.

CECA (October 2002). *Supply Chain Relationships*. Civil Engineering Contractors Association.

Construction Products Association (2000). *E-Construction: Are We Ready?* Computing Association.

Davis Langdon Consultancy (2002). *The Impact of E-business in UK Construction*. Department of Trade and Industry.

Douglas A. (2003). *Damage Limited*. Supply Management. 22 May.

Web sites

www.itcbp.org.uk
www.e-envoy.gov.uk
www.eu-supply.com
www.buyitnet.com
www.itw.itworld.com

Appendix

BMI STANDARD FORM OF PROPERTY OCCUPANCY COST ANALYSIS

BUILDING TYPE:		OWNER:

LOCATION:	DATE OF ERECTION:	OCCUPIER:

UPPER MANAGEMENT CRITERIA

BUDGET PROCEDURE

Estimate:

Budget:

Cost control:

MAINTENANCE MANAGEMENT AND OPERATION

Responsibility:

Total estate:

Routine inspections:

Painting frequencies:

Cost records and feedback:

Directly employed labour:

Trades:		Elec/Mech Fitters	Carpenters	Services Foreman	Premises Svc Asst	Premises Supervisor
Number:						

Incentive schemes:

Where are directly employed labour force is used ... % is added to basic labour rates and ... % to basic material costs to cover direct overheads.

Work done by DEL ... % and contracted-out ... %

Forms of contract:

Contract supervision:

BUILDING FUNCTION

Space use:

Number of occupants:

Design criteria:

Change of use:

FORM OF CONSTRUCTION

Structure:

External walls:

Windows:

Roof:

Internal partitions:

Floor structure:

Floor finishes:

Fittings & fixtures:

Internal decoration:

External decoration:

Plumbing:

Heating & ventilating:

Lifts & escalators:

Other M & E services:

PARAMETERS

Gross floor area:	Storeys above (& including) ground floor:	Height to ridge:
Area of pitched roofs (on plan):	Floors below ground floor:	Height to eaves:
Area of flat roofs (on plan):	Floor to ceiling height:	Area of external works:
Area of external glazing:		

Element	Total (£)		Cost per 100 m² floor area		Brief description of work
0. Improvements & adaptations		–		–	
1. Decoration 1.1 External decoration 1.2 Internal decoration **Sub-total**	——	£	——	– £	
2. Fabric 2.1 External walls 2.2 Roofs 2.3 Other structural items 2.4 Fittings & fixtures 2.5 Internal finishes **Sub-total**	——	£	——	– £	
3. Services 3.1 Plumbing & drainage 3.2 Heating & ventilating 3.3 Lifts & escalators 3.4 Electrical power & lighting 3.5 Other M & E services **Sub-total**	——	£	——	– £	
4. Cleaning 4.1 Windows 4.2 External surfaces 4.3 Internal **Sub-total**	——	£	——	– £	
5. Utilities 5.1 Gas 5.2 Electricity 5.3 Fuel oil 5.4 Solid fuel 5.5 Water rates 5.6 Effluents & drainage charges **Sub-total**	——	£	——	– £	

6. Administrative costs 6.1 Services attendants 6.2 Laundry 6.3 Porterage 6.4 Security 6.5 Rubbish disposal 6.6 Property management					
Sub-total	—— £		—— £		
7. Overheads 7.1 Property insurance 7.2 Rates					
Sub-total	—— £		—— £		
TOTAL	£		£		

External area ... m²	External works total (£)	Cost per 100 m² of external area	Brief description of work
8. External works 8.1 Repairs & decoration 8.2 External services 8.3 Cleaning 8.4 Gardening			
External works total	£	£	

Index

T - #0486 - 101024 - C0 - 234/156/18 - PB - 9780750658195 - Gloss Lamination